T0202309

Oxford's Savilian Professors of Geometry

S.R HENRY SAVILLE,
by Marcus Garrett.
From an Original Picture in the Bodleian Gallery Oxford.

Mathematics in Oxford from the 1620s onwards was dominated by the legacy of Sir Henry Savile (1549–1622). The professorships of geometry and astronomy that he founded in 1619 for 'persons of character and repute, from any part of Christendom, well skilled in mathematics and 26 years of age' were to have an unparalleled influence on the institutional structure of Oxford mathematics.

OXFORD'S SAVILIAN PROFESSORS OF GEOMETRY

The First 400 Years

Edited by

ROBIN WILSON

OXFORD
UNIVERSITY PRESS

OXFORD
UNIVERSITY PRESS

Great Clarendon Street, Oxford, OX2 6DP,
United Kingdom

Oxford University Press is a department of the University of Oxford.
It furthers the University's objective of excellence in research, scholarship,
and education by publishing worldwide. Oxford is a registered trade mark of
Oxford University Press in the UK and in certain other countries

Published in the United States of America by Oxford University Press
198 Madison Avenue, New York, NY 10016, United States of America

British Library Cataloguing in Publication Data

Data available

Library of Congress Control Number: 2021951696
ISBN 978–0–19–886903–0

DOI: 10.1093/oso/9780198869030.001.0001

Printed in Great Britain by
Bell & Bain Ltd., Glasgow

Sir Henry Savile, engraving by R. Clamp, 1796, after
S. Harding after M. Gheeraerts. Wellcome Collection.

FOREWORD

The year 2019 marked the 400th anniversary of the founding by Henry Savile of the Savilian chairs of geometry and astronomy at the University of Oxford. The British Society for the History of Mathematics held a meeting in November 2019 in Oxford's Bodleian Library to celebrate 400 years of Oxford's Savilian professors of geometry, whose number I had the great honour of joining a few years ago. I was delighted to take part in that celebration, from which this volume on the history of the Savilian professors of geometry has developed.

At the anniversary meeting I was asked to talk about the four 20th-century holders of the Savilian professorship of geometry since 1963: Michael Atiyah, Ioan James, Richard Taylor, and Nigel Hitchin. By coincidence I had met them all for the first time within a period of about twelve months, and Richard Taylor, the youngest of them, was the first. Richard and I were both undergraduates at Clare College in Cambridge – not at exactly the same time, but close enough to overlap. I was in my fourth year, taking Part III of the Mathematics Tripos, in 1980–81 when Richard was in his first year. It so happened that Clare's fellow in pure mathematics, S. J. (Paddy) Patterson, had just departed to Germany to become professor of mathematics at the University of Göttingen, and there was no replacement fellow in that year. So I was asked to give supervisions to the first-year undergraduates, including Richard; I remember him as a very impressive student.

At the time that I was teaching Richard, I was wondering what I might do after my Part III in Cambridge. I had enjoyed voluntary work with children as an undergraduate, so one possibility that appealed to me was primary school teaching, and I applied for a teacher training course at Durham University. My interviewers at Durham urged me to consider being a secondary school mathematics teacher, rather than a primary school teacher. On the train journey returning from Durham to Cambridge, I decided that if they really wanted secondary school teachers it made sense to pursue my other option first, which was a PhD degree in algebraic geometry. It had been suggested to me that I should contact Michael Atiyah for advice about this, and so he was the next of the four that I met.

I summoned up the courage to phone Michael on his home number. I think that it didn't occur to me to try to contact him at the Mathematical Institute, and electronic mail didn't exist at the start of the 1980s; I probably just looked up his name in the telephone directory. Although it was nearly lunchtime, it turned out that I had woken him up, as he had just returned from a trip to the United States. I was hugely embarrassed, but he was very kind and invited me to meet him, at which point he gave me a great deal of helpful advice and told me to put his name on my application form if I decided to apply to Oxford. This I did, and I had the enormous privilege of ending up as his DPhil student in the following autumn.

As a DPhil student in the Mathematical Institute, I shared offices with other students supervised by Michael. These included two other Michaels (one now my husband, who had been supervised initially by Ioan James, then Savilian professor of geometry) and also Simon Donaldson (who had been supervised initially by Nigel Hitchin). So I was soon aware of the presence in the department of both Ioan and Nigel, though of course to a lesser extent than that of Michael Atiyah. According to one of his nieces, Michael was known as the quiet member of his family, but that says more about his family than about Michael himself, who was typically to be found talking mathematics with loud exuberance to a colleague, student, or visitor somewhere in the Institute. However, I do remember that very occasionally he seemed stuck for words. When he came into the office I shared with Michael, Michael, and Simon, to summon one of us for a discussion, he would never address us by name but would just look at the relevant person. I have always wondered whether he felt awkward partly because we happened to share his own names (his middle name was Francis), except for Simon (who by chance has my surname as his middle name).

Michael Atiyah was without doubt one of the most influential mathematicians of the second half of the last century; as his former student and collaborator Graeme Segal has said, changing the (mathematical) landscape is what he will be remembered for. Very sadly, Michael died in January 2019, at the start of the anniversary year. I am sure that he would have enjoyed the anniversary celebrations and the opportunity to find out more about his predecessors as Savilian professors of geometry as much as I have done.

Frances Kirwan DBE, FRS
Savilian Professor of Geometry

CONTENTS

THE SAVILIAN PROFESSORS OF GEOMETRY

1619   **Henry Briggs** (1561–1631)

1631   **Peter Turner** (1586–1652)

1649   **John Wallis** (1616–1703)

1704   **Edmond Halley** (1656–1742)

1742   **Nathaniel Bliss** (1700–1764)

1765   **Joseph Betts** (1718–1766)

1766   **John Smith** (*c*.1721–1797)

1797   **Abraham Robertson** (1751–1826)

1810   **Stephen Peter Rigaud** (1774–1839)

1827   **Baden Powell** (1796–1860)

1861   **Henry John Stephen Smith** (1826–1883)

1883   **James Joseph Sylvester** (1814–1897)

1897   **William Esson** (1839–1916)

1920   **Godfrey Harold Hardy** (1877–1947)

1931   **Edward Charles Titchmarsh** (1899–1963)

1963   **Michael Francis Atiyah** (1929–2019)

1969   **Ioan MacKenzie James** (b. 1928)

1995   **Richard Lawrence Taylor** (b. 1962)

1997   **Nigel James Hitchin** (b. 1946)

2017   **Frances Clare Kirwan** (b. 1959)

Sir Henry Savile's monument in the antechapel of Merton College shows him flanked by Ptolemy and Euclid, the two foundational authorities whose work he expounded in his Oxford lectures.

CHAPTER I

Sir Henry Savile and the early professors

WILLIAM POOLE

This opening chapter addresses the state of mathematical instruction in Oxford, both before and after the foundation in 1619 of Sir Henry Savile's professorships in geometry and astronomy. After discussing the terms and conditions of his foundations, we situate Savile's benefaction within the general context of Oxford teaching and learning structures. Savile personally appointed his first two professors, Henry Briggs and John Bainbridge, and this chapter discusses their performance and publications, asking what changed with their appointments. Finally, words are offered on their Savilian successors, Peter Turner and John Greaves.

The mathematical scene

In 1619 the scholar and statesman Sir Henry Savile established his two professorships for the University of Oxford, one in astronomy and the other in geometry. It was a great age in that university for charitable endowments, and these were the earliest university chairs in England in the mathematical arts.[1]

They were not entirely novel, however. For a nearby prompt, we need look only to London and the foundation of the professorships at Gresham College, active from 1597

William Poole, *Sir Henry Savile and the early professors*. In: *Oxford's Savilian Professors of Geometry.*
Edited by Robin Wilson, Oxford University Press. © Oxford University Press (2022). DOI: 10.1093/oso/9780198869030.003.0001

and covering seven subjects, including geometry and astronomy. Savile, indeed, would poach his first geometry professor from the Gresham chair. But then, as now, Gresham College neither enrolled students nor awarded degrees.

North of the border, in Scotland, there had been a professor of mathematics at St Mary's College, St Andrews, from 1574. It was moved by Act of Parliament to St Salvator's College in 1579, but the position was not endowed, and lapsed following the death of its incumbent in 1603.[2] The professorship in mathematics at Edinburgh's university (the 'Tounis College') dates from 1620, when one of the four 'regents' in philosophy was given additional duties and salary as the 'public professor of the mathematics', but this got off to a faltering start.[3] In 1612 at Aberdeen, however, the mathematician, astronomer, and royal physician Duncan Liddel (1561–1613) had created six bursaries in arts and mathematics for his *alma mater*, Marischal College, and in the following year he endowed a permanent professorship there in mathematics.[4] This is perhaps the closest earlier British analogue to Savile's benefaction, for Liddel also gave to the college his library, along with an income to purchase annually 'new books of most ancient mathematicks globis and instruments'. But once again, successfully filling the post proved difficult, and Savile's chairs were occupied first.

Henry Savile, however, was not a man to admit precedent. When he resolved upon the two professorships that bear his name, like many a benefactor before and since he found it convenient to exaggerate the problem that he was so munificently to fix. The mathematical arts, he claimed in his deed of foundation, were 'a quarter almost given up in despair', 'uncultivated by our countrymen';[5] and in truth, the Oxford system at least could benefit from his precise intervention. Undergraduates across Europe had in theory always studied a little mathematics: the earliest statutes of Oxford, for instance, mentioned no fewer than five prescribed topics – 'Geometria' (geometry), 'Algorismus' (counting), 'Spera' (the sphere), 'Compotus' (calendrical calculation), and 'Arsmetrica' (measuring).[6] By the late 16th century, the teaching of the mathematical arts in the English universities rested on two complementary institutions: the general college tutor, and the recent MAs, known in Oxford as 'regent masters', several of whom were employed to give public lectures. The problem was that the college tutors might possess only the most basic knowledge, held over from their own undergraduate days, and the regent masters might be little better – it is not so very different from the modern practice of farming out teaching to new graduates.

This should not blind us to the possibility that remarkable individuals could still be produced by this fairly minimal regime. The most notable Oxford example in the earlier Elizabethan period was Thomas Allen (BA, Trinity College, 1563; MA, Gloucester

(*Left*) Thomas Allen (1542–1632) of Gloucester Hall (later refounded as Worcester College) was a collector of mathematics books and manuscripts, of which many came to the Bodleian Library.
(*Right*) Erasmus Williams (*fl.* 1590) of New College was a versatile scholar who also mapped College lands. This memorial brass in the Church of St Mary Magdalene, Tingewick, illustrates his interests, including music, astronomy, and geometry.

Hall, 1567), who from the 1560s built up an unparalleled collection of mathematical manuscripts and instruments and taught many a younger mathematician of future note. When he died in his 90s in 1632, his eulogist, William Burton, claimed that when Allen first delivered his required MA lectures in mathematics he had feared that, with the multitude of learned men in attendance, 'the very walls might burst'; Burton then intriguingly contrasted this sharply with the situation in the 1630s.[7]

In the generation after Allen, Thomas Lydiat of New College was one of the few Oxonian mathematicians to be noticed internationally. He specialized in the interrelated fields of chronology and astronomy and, although a biblical literalist and geocentrist, he was happy to propound, before Kepler, the possibility of oval (rather than perfectly circular) planetary orbits. Furthermore, the level of mathematical interest in the later Elizabethan and Jacobean periods can be gauged by book ownership, and here we note

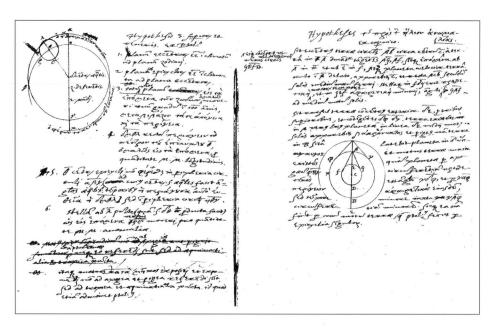

Savile's lectures, the earliest public teaching in Oxford of the new astronomy for which we have evidence, presented both the Ptolemaic (*left*) and Copernican (*right*) models for the Sun's apparent motion.

Erasmus Williams, also of New College, who left his copy of Copernicus's *De Revolutionibus Orbium Coelestium* (On the Revolutions of the Heavenly Spheres) to the College library in his will; it is still there. Finally, there was Savile himself, from Merton College, whose own mathematical lectures as a regent master were (like Allen's) celebrated. He expounded both Euclid and Ptolemy, and also Copernicus.

Lydiat's first published work, the *Prælectio Astronomica* of 1605, derived from lectures given in his college around the turn of the century, and this alerts us to the third development in the time of Savile, the rise of the specialist college lecturer. If the regency system was breaking down – students could pay not to attend – and if the colleges were having trouble fielding adequate general tutors, then perhaps it was time to invent the college-based specialist. Accordingly, by the time that Savile founded his chairs, around half a dozen Oxford colleges were employing bespoke lecturers in mathematics, alongside similar posts in the more traditional subjects, and in some rarer cases, such as at Merton College, in modern languages too.

So, while the colleges were slowly evolving from general to specialist tutors in the mathematical arts, the University itself was stuck on the regency system. Despite the Regius professorships in divinity, Greek, and Hebrew, there was no corresponding foundation for mathematics. Just as the colleges were specializing, so the University needed to do so too; this is the intervention which Savile made, and this is its historical significance. In the words of Seth Ward, the Savilian Professor of Astronomy, in 1654, mathematics was now 'faithfully expounded in the Schooles by the professor of geometry, and in several Colledges by particular Tutors'.[8]

Henry Savile

Henry Savile was a Yorkshireman with a learned father and learned brothers, and was educated from the age of 12 at Brasenose College and then at Merton College, where he became a fellow. He took his MA degree in 1570 and was appointed one of the regent masters in astronomy for the following academic year. His own account of his mathematical development at this time is significant: he started on Euclid's *Elements*, the curricular staple, but switched to Ptolemy's *Almagest* when meeting increasing difficulty in the middle of Euclid's tenth book. Ptolemy in Greek then proved equally hard, and so Savile went back to Euclid, finished it, and returned to complete Ptolemy.

There may be an element of posturing here, for, as a modern authority on Savile has commented,[9] Savile may have been advertising his allegiance to Platonic thought: Plato in his *Republic* had recommended for his ideal citizens a similar plan of study in the mathematical arts, starting with arithmetic, moving from plane to solid geometry, then to astronomy, and finally to harmonics. For Savile was above all a humanist scholar, a student of classical antiquity. Such scholars were interested in three things: producing better editions of surviving texts, establishing the history of their discipline, and recovering lost texts. Accordingly, Savile studied Ptolemy in effect by editing him, and his partial translation into Latin survives among his manuscripts in the Savile collection in Oxford's Bodleian Library. As for the history of his discipline, Savile prepared an ornate biobibliography of all ancient mathematicians, again in a work that he left in manuscript.

Finding new texts was usually facilitated by scholarly travel, and in 1578 Savile, supported by Merton College, set off on a Continental tour that lasted for four years. While travelling, he made friends with several leading mathematical humanists, including Johannes Praetorius in Altdorf, Thaddaeus Hagecius in Prague, and Andreas Dudith and Paul Wittich in Wrocław. During his tour, Savile also made copies of rare manuscripts – for instance, the then unknown and unpublished *Isagoge* of the Hellenistic astronomer Geminus of Rhodes. In Vienna, Savile was pointed to a copy of this text, which he transcribed with his travelling companion George Carew; it is now among the Savilian manuscripts.[10] Savile then moved on to Padua to enjoy the collections and company of Gian Vincenzo Pinelli, humanist, mentor of Galileo, and the major 16th-century Italian collector of printed books and manuscripts. While there, further copies were taken from Savile's copy of Geminus by Pinelli's own scribe.

Similar activities on this Continental tour account for several other items today in the Savile collection and elsewhere. We will return to this humanist emphasis, because Savile never really lost it. Nor is it in any way incompatible with Savile's attested interest in Copernicus, who himself advertised his own system as restoring an ancient Pythagorean tradition.

Savile's foundation

Savile's international reputation as a bible translator, a patristic editor, and a scholar and translator of Roman history is not of concern here, but as mathematics was his first love, so it was his last. Instead of founding chairs in his other passions, such as Greek or Divinity, he chose instead geometry and astronomy.

What motivated him? As we stated at the outset, this was an age of benefactors and, as in so many other things, Savile was led here by his friend and older Merton College contemporary Sir Thomas Bodley, the development of whose library, which had opened in 1602, Savile helped to steer after Bodley's death in 1613. If we look at the major endowments in Oxford at this time, in chronological order they comprise Bodley and the Bodleian Library (1602), the Sedleian chair of natural philosophy (bequest of 1618, active from 1621), the Savilian chairs of geometry and astronomy (1619), White's chair of moral philosophy (1621), the Earl of Danby's Botanic or 'Physic' Garden (1621), the Camden professorship of history (1622), the merchant Richard Tomlins's readership in anatomy (1624), the Heather professorship of music (1626), and the Laudian professorship of Arabic (1636). Thomas Bodley lacked a direct male heir, and the bulk of his fortune went to the University. Henry Savile also lacked a direct male heir, as did Thomas White, who had endowed by testament his chair of Moral Philosophy; and William Camden never married.[11] These men were perhaps more focused on legacy than they might otherwise have been.

What is also noticeable is the growing specialization of such benefactions, as areas in the chiefly traditional 'arts' curriculum started to be perceived as requiring expert handling at the University level. By Savile's time, the University had no need for professorships in Greek or divinity – it already had such posts – and to make an impact, Savile turned to tidying up the situation in the precise subjects in which he could justly claim status as the English authority.

Thomas Bodley's Schools quadrangle was built between 1613 and 1624. The pillars decorating the central tower correspond to the five orders of architecture. On the first floor the Savilian Professor of Astronomy lectured to the north of the central tower, while the Savilian Professor of Geometry lectured to the south. The tower itself was fitted out by John Bainbridge, the first Savilian Professor of Astronomy, as an astronomical observatory.

Henry Savile's foundation was distinguished in several ways. First, his professors were exceptionally well salaried on £160 per annum, the income deriving from lands settled on the University for the purpose. Real income fluctuated with rents, but this statutory income was about three times the basic salary of a Head of House, and fifteen times that of an ordinary Fellow, before augmentations for other College offices. Savile also provided a triple-locked mathematical chest, a bank from which his professors, with University permission, could draw for costs such as the purchase of instruments.

To protect their time for teaching and research, Savile also barred his professors from senior academic or ecclesiastical office. Although, as we shall see, the professors were expected to offer tuition at all levels, their focus, in teaching as in research, was to be on advanced mathematics, it being assumed that College tuition would have furnished the basics. Savile's next innovation was to furnish his professors with a library, complete with a set of mathematical instruments. As he wrote to William Camden when advising him on how to set up his own professorship:[12]

(*Left*) 'Cista mathematica', Henry Savile's chest, was part of his provision for the study of the mathematical sciences at Oxford.
(*Right*) An alabaster model associating the five classical orders of architecture with the five Platonic solids, probably used in the 17th century for geometry teaching.

One thing more I will be bold to persuade you, that to the use of your Readers you would bequeath your Books of that faculty. I for my part have cleared my study of all the Mathematical Books, which I had gathered in so many years and Countreys, Greek and Latin, printed and manuscripts, even to the very raw Notes, that I have ever made in that argument.

This library is the golden thread that ran through the professional lives of the earliest professors, and we would be justified in seeing many of the major editorial projects undertaken by the professors in this time as intergenerational collaborations – the Oxford 'Euclid' of 1703, for instance, rested on a chain of editorial work that reached back to Savile's own 16th-century annotations.[13]

The next point of interest is Savile's stipulations for lectures. Sticking to the major texts for now, the geometer was required to concentrate on Euclid. The astronomer was to teach Ptolemy, alongside Copernicus and others if he wished, and also (for introductory purposes) 'the sphere of Proclus, or Ptolemy's hypotheses of the planets'. This is a significant detail, to which we shall return.

Finally, there is the issue of 'experimentalism', or the place of practical work under the earliest Savilian professors. Today the professorships could not be more different: the current geometry professor usually works in areas of mental abstraction that are closed to non-specialists, but with the ironic complication that computing now plays a role that Savile could never have imagined; the current astronomy professor deals with the Sun's internal rotation and with telescope arrays that are trailed in space to form vast triangles inscribed in planetary orbits.

So what did Savile say about the relationship between the theoretical and the practical? The evidence here is mixed, and perhaps the best solution is to agree that Savile warmed over time to the importance of practical science, but that he was not himself particularly drawn towards it. Savile's last lectures, on Euclid, were humanist to the final sentence and were concerned more with the history of mathematics – establishing for the first time which of the two historical Euclids (Euclid of Alexandria, rather than Euclid of Megara) was the mathematical author, and continuing with the explication of theorems, rather than with instrumentation or any real-world application.

But despite his notorious dismissal of his first choice for the Savilian Professor of Geometry (described below), Savile's own statutes show that he had come to believe that mathematical instruction ought to include a knowledge of practical or 'mixed'

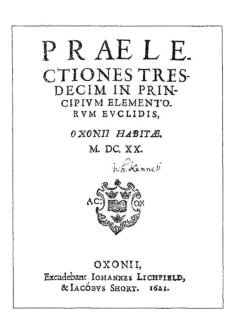

Savile's introductory lectures on Euclid's *Elements* were published in 1621.

mathematics. Among his other duties, the geometer was required to teach land-surveying, while the astronomer had to cover optics, gnomonics, geography, and 'the rules of navigation in so far as they are dependent on mathematics'. The geometer was to lead his students to the fields, and the astronomer was to carry out, record, and archive his observations. Savile even permitted his geometer to offer the most basic instruction in English, a remarkable concession for the time.[14]

Henry Briggs, Savile's first geometer

Savile appointed his first two professors personally. For his geometrician he eventually chose a fellow Yorkshireman, Henry Briggs.[15] Savile delivered the first geometry lectures of his new institution, commencing ceremonially in the Divinity School in Trinity Term 1620, before moving into the new Geometry School at some point in the following Michaelmas Term. At the end of 1620, Savile concluded a series of lectures on the first eight propositions of Book I of Euclid's *Elements* by saying:[16]

Trado lampadem successori meo (I hand the torch to my successor).

So on 8 January 1621, Briggs dutifully took up the torch, starting with the ninth proposition.

Yet Briggs had not been Savile's first choice. A much-repeated anecdote about Savile's foundation concerns Edmund Gunter of London, Savile's initial interviewee:[17]

he first sent for Mr Gunter, from London (being of Oxford University) to have been his Professor of Geometrie; so he came and brought with him his Sector and Quadrant, and fell to resolving of Triangles and doeing a great many fine things. Said the grave Knight [Savile], *Doe you call this reading of Geometrie? This is shewing of tricks, man!* and so dismisst him with scorne, and sent for Henry Briggs, from Cambridge.

This splendid account – it is from John Aubrey and is unverifiable – might be taken simply to mean that Savile still preferred a University theoretician to a metropolitan manipulator of instruments, but some qualifications need to be made.

Briggs, 'from Cambridge' (in terms of his earlier education) was at that point Gresham Professor of Geometry in London – and Gunter was an Oxonian and a senior theologian by academic qualification with a BD degree from Christ Church in 1615. In London, Gunter was close friends with Briggs, who had mentored him and suggested him as successor to the first Gresham Professor of Astronomy (Edward Brerewood), a post that he eventually obtained around the time that he was failing to impress Savile. Furthermore,

Briggs was also devoted to the 'resolving of Triangles' and similar practical operations, as we shall see, so the supposed opposition here between the city instrumentalist versus the ivory tower theorist dissolves upon inspection.[18] But for all that, Savile's core bias is apparent. Mixed mathematics might have its uses, but pure mathematics (then synonymous with 'geometry') was still the higher science.

Henry Briggs had been educated at St John's College, Cambridge, gaining his BA degree in 1582 and his MA in 1585. There, 'drawn by a certain propensity of nature', he devoted himself to mathematics, and 'not merely stuck at the bark and surface but penetrated into the marrow and more hidden secrets of those sciences'.[19] His biographer mentions no formal instruction, but Briggs may have been taught by John Fletcher of Gonville and Caius College, who was celebrated for his mathematical ability. Briggs was appointed Cambridge University's Mathematical Lecturer for the year 1587–88 and wrote some commentaries on the geometry of Ramus in manuscript. Bearing the date 1588, these have been identified in modern times and illustrate the kind of mathematics that was offered at university level at that time.[20]

At St John's College Briggs was appointed in mid-1592 to the position of *Mathematicus Examinator* (mathematical examiner). This was a lectureship established by the 1524 statutes of the College, and in theory was held simultaneously by four fellows who were to offer daily lectures over the long vacation for the *Baccalaurei* (the BA students who were reading for their MA degree) in arithmetic, geometry, perspective, and the sphere or cosmography. From the next term he was also appointed to the College's much better paid Linacre lectureship in medicine. Judging from the College accounts, however, it seems as if there was only one mathematical *examinator* by Briggs's time, with the others handling logic and philosophy. But as Briggs was paid 10s quarterly, like the other *examinatores*, it seems that one mathematical lecturer now covered all four areas, spaced throughout the academic year.[21]

Briggs certainly taught the young Thomas Gataker who received his BA degree in 1594 and his MA degree in 1597, and who later became a famous scholar.[22] This willingness to offer collegiate instruction remained with Briggs, for when he moved to Oxford over three decades later, he taught arithmetic three times a week to the undergraduates of his new society, Merton College, independently of his Savilian duties.[23] Briggs would happily have remained in Cambridge in the 'shades of the academe' (said his first biographer), had he not been drawn by divine providence to London, the 'theatre of the world'.[24] He did not forget his old College and University, however, and one of his first concerns upon election to the Savilian chair in Oxford was to offer any assistance that he could towards the foundation of a parallel chair at Cambridge.[25]

In 1597 Briggs became the first Professor of Geometry at Gresham College in London, with rooms and a salary of £50 per annum.[26] Savile's later statutes may owe something to the Gresham provisions, for there both the geometer and astronomer were to lecture in Latin in the morning on their appointed day, but in English in the afternoon, and both courses were split throughout the year between theoretical and practical topics.[27] Briggs's earliest publications were all short contributions to other men's works, exclusively comprising tables with the odd explanatory text, and all of them concerned navigation.[28] But this was not really the kind of mathematics where the mathematical humanist Henry Savile's heart lay.

Moreover, on the evidence we have, Briggs spent a great deal of his time engaged in mathematical astronomy. In 1609 he became friends with the great scholar James Ussher, at that point a theologian and churchman in Dublin, and it is deeply regrettable that all but two letters from their apparently extensive correspondence is lost. Their exchanges concentrated on astronomy and chronology, and it was during this period that Briggs set about computing improved eclipse tables.[29] For this purpose he started reading Kepler's works avidly, but with some frustration as he was unsympathetic towards Kepler's cosmology, intolerant of his astrology, and uninterested in his Platonism. Indeed, those who study Kepler's reception in England during this period rightly judge Briggs to have been drawn chiefly by the computational possibilities of Kepler's work. On the slim surviving evidence, Briggs preferred a non-realist position in which, although content with the basic Copernican assumptions, he rejected Kepler's ellipses, electing instead to work on older circular models.[30] He retained his reputation for both geometry and astronomy until his death, and after: a poetic epitaph in Greek on him by the Merton scholar and poet Henry Jacob hailed him as:[31]

βριγγίαδης ζωστήρ γαίης, καὶ σύνδρομος ἄστρων,
᾽Ευκλείδεν φρονέων, καὶ Πτολεμαῖον ὅλους.
(Briggs, the land's girdle, and meeting-place of the stars, mindful of Euclid, and all Ptolemy.)

It is in the latter of Briggs's two surviving letters to Ussher, from March 1615, that we first hear of a new excitement in his scholarly life. For Briggs changed mathematical direction when he discovered the work of the Scottish mathematician John Napier, on logarithms. The logarithm (a Greek word, coined by Napier, meaning 'ratio-number') is a computational shortcut that replaces multiplication by addition, as the sum of the logarithms of two or more numbers is the logarithm of their product. This he encountered in Napier's recent publication on the subject, the *Mirifici Logarithmorum Canonis Descriptio* (A Description of the Marvellous Table of Logarithms) of 1614.

MIRIFICI

Logarithmorum
Canonis deſcriptio,

Ejuſque uſus, in utraque
Trigonometria; ut etiam in
omni Logiſtica Mathematica,
Ampliſſimi, Faciliimè, &
expeditiſſimi explicatio.

Authore ac Inventore,
IOANNE NEPERO,
Barone Merchiſtonii,
&c. Scoto.

EDINBURGI,
Ex officinâ ANDREÆ HART
Bibliopôlæ, cIↃ. Dc. xiv.

John Napier's *Descriptio* of 1614.

Briggs, who became obsessed almost to exclusion with Napier's discovery, started lecturing at Gresham College on logarithms, suggested an improved system, wrote to Napier about it, and visited the Scots nobleman at his seat in Merchiston Castle in the summer of 1615 when Briggs had no lecturing duties; he visited again in the next summer, and was frustrated in a third visit only by Napier's death in April 1617. The men evidently got on well – indeed, when they first met, 'almost one quarter of an hour was spent, each beholding other almost with Admiration before one word was spoke'[32] – and Briggs is to be credited with either initiating or consolidating Napier's shift from a variant of natural logarithms to logarithms that were calculated to base 10.

Logarithms define and exploit a connection between arithmetic and geometric progressions. This pleasing idea is a theoretical one, and was laid out in that way by Napier, but astronomers and navigators (their terrestrial equivalents) now had a powerful computational tool for solving trigonometrical problems – precisely the 'resolving of Triangles' for which Gunter had been so mocked by Savile. Whereas Napier's conception of logarithms was kinematic (points moving in space), Briggs was solely interested in numerical techniques – the two men thought about number concepts rather differently.[33] Some sense of Briggs's assumptions about the application of logarithms is given by his comment to Ussher, before he had even opened communications with Napier, that

Logarithmi.

1	0,0000,00000,00000
2	0,3010,29995,66398
3	0,4771,21254,71966
4	0,6020,59991,32796
5	0,6989,70004,33602
6	0,7781,51250,38364
7	0,8450,98040,01426
8	0,9030,89986,99194
9	0,9542,42509,43932
10	1,0000,00000,00000

Num. absolut.	Logarithmi.	Num. absolut.	Logarithmi.	Num. absolut.	Logarithmi.
17501	4,24306,28648,0481	17534	4,24388,10022,1832	17567	4,24469,76012,9672
	2,48147,0057		2,47679,9920		2,47214,7329
17502	4,24308,76795,0538	17535	4,24390,57702,1752	17568	4,24472,23217,7001
	2,48132,8279		2,47665,8675		2,47200,6614
17503	4,24311,24927,8817	17536	4,24393,05368,0427	17569	4,24474,70428,3615
	2,48118,6517		2,47651,7447		2,47186,5916
17504	4,24313,73046,5334	17537	4,24395,53019,7874	17570	4,24477,17614,9531
	2,48104,4771		2,47637,6234		2,47172,5233
17505	4,24316,21151,0105	17538	4,24398,00657,4108	17571	4,24479,64787,4764
	2,48090,1041		2,47623,5037		2,47158,4566
17506	4,24318,69241,3147	17539	4,24400,48280,9145	17572	4,24482,11945,9330
	2,48076,1329		2,47609,3857		2,47144,3915
17507	4,24321,17317,4476	17540	4,24402,95890,3002	17573	4,24484,59090,3245
	2,48061,9631		2,47595,1692		2,47130,3280
17508	4,24323,65379,4107	17541	4,24405,43485,5694	17574	4,24487,06210,6525
	2,48047,7951		2,47581,1544		2,47116,1661
17509	4,24326,23427,2058	17542	4,24407,91066,7238	17575	4,24489,53336,9187
	2,48033,6186		2,47567,0411		2,47102,1059
17510	4,24328,61460,8344	17543	4,24410,38633,7650	17576	4,24492,00439,2146
	2,48019,4638		2,47552,9296		2,47088,1473
17511	4,24331,09480,2982	17544	4,24412,86186,6946	17577	4,24494,47527,2719
	2,48005,3005		2,47538,8196		2,47074,0902
17512	4,24333,57485,5987	17545	4,24415,33725,5142	17578	4,24496,94601,3621
	2,47991,1389		2,47524,7112		2,47060,0347

(*Left*) Some logarithms from Briggs's *Logarithmorum Chilias Prima*, calculated to 14 decimal places (1617).
(*Right*) Some logarithms from Briggs's *Arithmetica Logarithmica* of 1624.

'I purpose to discourse with him concerning Eclipses, for what is there which we may not hope for at his hands'.

Almost all of Briggs's subsequent publications were in the field of logarithms. After first supplying a table for Edward Wright's translation of Napier's *Descriptio*,[34] Briggs published a 16-page pamphlet, entitled the *Logarithmorum Chilias Prima*, which was an initial thousand (or 'chiliad') of his own logarithms to base 10. Tellingly, one of the only four known surviving copies is bound with the Gresham Professor of Astronomy's tables of sines and tangents, Edmund Gunter's *Canon Triangulorum* of 1620, evincing once again the close connection between logarithms and the trigonometry essential for mathematical astronomers and geographers.[35]

In 1624, after years of grinding calculation, Briggs brought to the press his *Arithmetica Logarithmica*, tables of logarithms of numbers from 1 to 20,000 and from 90,000 to 100,000, calculated by hand to 14 decimal places. His own copy has parchment finger-tabs inserted for every thousand numbers, and a few further copies have an extra chiliad at the end.[36] The gap from 20,000 to 90,000 was filled by Adriaen Vlacq of the Netherlands, and the complete set was published in Gouda in 1627, in Latin and French editions, but extending to only ten decimal places. Various translations and vulgarizations followed, but Briggs's remaining mathematical writings were published posthumously by his disciple Henry Gellibrand and printed again in Gouda in 1632 by an associate of Vlacq as the *Trigonometria Britannica*.[37] The Briggs–Vlacq tables had a long and wide afterlife, of which perhaps the most striking instance was afforded by two Chinese editions of 1713 and 1721.

Logarithms simplified trigonometry, trigonometry assisted navigation, navigation promised trade, trade yielded profit, and Henry Briggs carried his interests along that chain. He was a member of (and a significant shareholder in) the Virginia Company, in whose Court Book he appeared regularly from April 1619 to June 1620, being about to leave London for Oxford and resigning his Gresham professorship in July of that year.[38] In 1622, after he had taken up his new position, Briggs's short English essay 'A treatise of the northwest passage to the South Sea, through the continent of Virginia and by Fretum Hudson' was published as an appendix to the colonist Edward Waterhouse's *A Declaration of the State of the Colony and Affaires in Virginia*. It was then republished in Samuel Purchas's famous collection of travel documents in 1625, this time with a map of 'The North part of AMERICA' signed by Briggs.[39] His entrepreneurship persisted: John Aubrey reported (but did not date) a scheme mooted by Briggs to connect the River Thames to the Avon by an artificial canal, and in 1630 Briggs was involved in a bid with Cornelius Drebbel and others to drain the East Anglian counties.[40]

By the time that Henry Briggs was appointed to the Savilian chair, he had become the obvious choice, despite Aubrey's story about Edmund Gunter. He had both academic and metropolitan teaching credentials, an international correspondence, attested interests in astronomy and geometry and in geography and navigation, and he was the prime mover in England of the new art of logarithms. At Oxford he quickly made himself useful within the University, sitting (for instance) on the committee appointed to oversee the newly founded Botanic Garden.[41] He maintained his overseas correspondence, and despite his strong (even puritan) Protestantism he assisted Roman Catholic scholars abroad, for in 1625 he provided *lectiones variantes* taken from the rare transcripts of Geminus to be found among the Savilian manuscripts. Both Henry Briggs and his Savilian colleague John Bainbridge were thanked warmly from Paris by the Catholic convert and librarian Lucas Holstenius, to whom Briggs also wrote in 1626 about his ongoing plans to complete his edition of Euclid's *Elements*.[42] In the late 1620s, Briggs corresponded at length with the Danish mathematician Longomontanus on the latter's doomed attempts to square the circle.[43] He also wrote a short paper in Latin on mathematical discoveries unknown to the ancients for the 1630 edition of a book by his Exeter College colleague George Hakewill. This book, *An Apologie for the Power and Providence of God*, was a kind of collective Oxford statement on modern intellectual achievement,[44] and Briggs headed it with two pieces of astronomy: the Copernican hypothesis, 'multo facilius & accuratius' (much simpler and more accurate) than the Ptolemaic model, and Galileo's discovery of the moons of Jupiter. These were then followed by algebraic and geometrical discoveries, including those of Reinhold and Regiomontanus, Harriot's method of finding the area

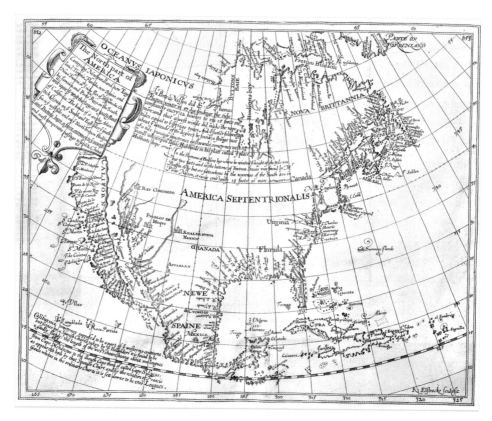

Henry Briggs's map of North America in *Purchas His Pilgrimes*.

of various shapes, and Napier's logarithms.[45] Again, Briggs assumed the provinces of both geometry and astronomy, with the latter (as ever) under its mathematical aspect, whereas Bainbridge's contribution to the same book provided a short paper in English on the 'supposed removeall of the Sun', a learned piece on the history of such observations that was more oriented to astronomy in its physical aspect.[46]

Briggs's commitment to the practicalities of mathematics is well illustrated by a letter that he wrote in 1626 to Thomas Lydiat, who had proposed a new period of 592 years as a tool for correlating world chronologies. Briggs was politely impressed, but showed himself in favour of persisting with era measures that were either of long standing or widely held, and this led him to remark that the time had finally come to adopt the Gregorian calendar, 'so that trade throughout the whole world may be the more easily plied by merchants' ('ut commercia per universum orbem a mercatoribus commodius exercerentur'). This continued to prove unpalatable to the British for over a century, but Briggs was pointedly aligning himself with the views of John Dee, whose similar proposal the government had charged a committee (which included Henry Savile) with investigating back in the 1580s.[47]

As a Savilian, Henry Briggs evidently lectured conscientiously, and he was a genial supporter of younger mathematicians such as John Pell, both in person and by correspondence.[48] But after his Oxford appointment, his only major publication that was not connected with logarithms was a partial edition of Euclid's *Elements* in 1620, comprising the first six books in Greek with a parallel Latin translation that was a corrected version of the standard one by Commandinus. Briggs's first biographer praised this edition, and it was evidently widely used, as the many copies extant today (often with rather distinguished provenances) attest. But it is a spartan and limited affair, published anonymously, launching straight into the text without prefatory remarks and featuring no commentary or explanations of its textual basis or rationale; moreover, the printed title page makes no mention that only six of the promised thirteen books were to be found there. But Briggs planned to complete this edition, and elsewhere he promised a text collated against two manuscripts and Savile's own adversarial copy. However, he could not find a publisher to take it on, a reminder to us that it was economically difficult to publish mathematical books at this time.[49] Nevertheless, it is a symbolically important publication, for reasons that will become apparent when we come to Briggs's Savilian colleague, John Bainbridge.

When Briggs died, he had set aside by testament some books for the precious Savilian library which he shared and augmented with Bainbridge. These comprise a telling mix of the ancient and the modern, the geographical, mathematical, and chronological: 'Ptolemie his geographie', 'parte of Mercator his great Atlas', 'Petavius de emendatione temporum', and his own works on logarithms with the Vlacq continuation – all remain there today. His testament also demonstrates his continued commitment to navigational discovery: £10 was directed 'towards the discovery of the Northwest passage'.[50]

We conclude our discussion of Briggs's career with a general remark about what changed in Oxford teaching and learning with his appointment. Oxford's regent masters had been young and untested: those formally involved in mathematical instruction were what we would today call 'early career academics'. Most of them would not go on to pursue a mathematical career, because there were then no careers in mathematics to pursue. By the time that Briggs took up his post he was in his late 50s and had already held the only other comparable job in the country. He was again, as we would say, 'a field leader' and held his job in perpetuity from year to year. There was no retirement in this period, only resignation or death. Briggs was able to bring to the job not just his skill,

Henry Briggs's memorial in the antechapel of Merton College.

but also his reputation and his contacts – as we have seen, his overseas correspondents included Kepler, Hondius, and Longomontanus – and it was only with this stability and stature that a subject could successfully 'bed in'.

John Bainbridge, Savile's first astronomer

Briggs's colleague was the younger man John Bainbridge, from Leicestershire.[51] As our focus is on the geometers we comment more briefly on Bainbridge, but we must recall that there was considerable fluidity between the two professorships in the earlier period, in terms of both province and personnel, and a study of either chair in isolation is no history at all.

Like Briggs, Bainbridge had been educated at Cambridge, at Emmanuel College, and like Briggs he also studied medicine, which he pursued to doctoral level and into actual practice. Just as Briggs had performed as Linacre lecturer at St John's College, so did Bainbridge read the corresponding Linacre lecture in Merton College several decades later, simultaneously with his role as Savilian Professor of Astronomy. He was still in his 30s when appointed to the Savilian chair and had just published *An Astronomicall Description of the Late Comet* (1618/19), a work that displayed some tolerance for astrology. This (possibly superficial) interest he would quickly drop and, perhaps not unconnected to this, judicial astrology was despised by both Savile and Briggs – Savile, indeed, had banned its teaching.[52]

Following their appointments, Bainbridge joined Briggs at Merton College, and in Oxford he pursued his twin careers as an astronomer and as a physician, having taken his medical doctorate in Cambridge in 1614. He commenced his duties with a public lecture that was held on the day following Briggs's own inaugural speech.[53] The two men's interests often overlapped: we have seen that Briggs wrote to Ussher to say that he had

John Bainbridge (1582–1643), the first Savilian Professor of Astronomy.

intended to construct better eclipse tables, and Bainbridge's earliest surviving letter to the same recipient promised exactly the same product.[54] According to John Aubrey, it was the geometer and not the astronomer who constructed the Merton College sundial;[55] but Bainbridge greatly augmented the Savilian stock of instruments, of which a catalogue and a deed of gift in his hand survive among his papers in Dublin.

John Bainbridge was underpublished during his lifetime, but like Henry Briggs he had international contacts, corresponding with the mathematicians and orientalists Willebrord Snellius, Thomas Erpenius, and Jacobus Golius (all in Leiden) and Martinius Hortensius (in Amsterdam). In Paris, the astronomer Ismaël Boulliau received and published some eclipse observations made by Bainbridge at Oxford in 1628 and 1630, and in 1631 Bainbridge even attempted to orchestrate an astronomical expedition to South America.[56] But if we turn to his first publication in post, we observe something very interesting. Savile's geometer was by statute to lecture on Euclid; this is why Briggs published his partial edition of Euclid's *Elements* in 1620 – to honour his terms of appointment through the humanist activity of editing the text on which he was to lecture. As for Savile's astronomer, he was instructed to lecture on Ptolemy's *Almagest*, but also on the shorter introductory texts 'the sphere of Proclus, or Ptolemy's hypotheses of the planets'.[57] Now in the same year as Briggs's Euclid, and with the same publisher, Bainbridge produced an edition of Proclus's *Sphaera*, followed by Ptolemy's *Planetary Hypotheses*. His preface, which eloquently expresses the duty of the professors to rescue and repair the texts of their ancient predecessors, shows that he had intended to edit Ptolemy's *Almagest*, just as Briggs had gone straight to Euclid, but that this had proved too great a task and so Bainbridge had begun with these simpler and shorter texts as harbingers for the editorial feat to come.[58]

This was a wise decision, but it was more profound than its modesty reveals. For the classic *Sphaera* of Proclus, a medieval evergreen, was actually not written by Proclus, but was rather a translated extract from the Hellenistic astronomer Geminus, encountered earlier, whose first edition had appeared only in 1590 in Altdorf from a text supplied ultimately by Henry Savile himself from his Viennese copy.

As for Ptolemy's *Planetary Hypotheses*, this was the first ever (if incomplete) edition of what had hitherto been a lost text. This short work, even in its drastically truncated form, served as a simple introduction to Ptolemy's world system up to the orbit of Saturn. Bainbridge's preface shows that he recognized the true nature of his former text as an excerpt from Geminus, and that he had found multiple copies of both of his texts among Savile's own manuscripts that were now in the professorial library. This double edition, with Ptolemy's chronological *Canon* as a short appendix, has been much commended, and is a more significant intervention than Briggs's cleaned-up republication of the first six books of Euclid's *Elements*. But when we pair these acts of publication, their true humanistic significance becomes fully visible.

Bainbridge continued Savile's humanist vision in other ways too. As we have seen, one of Savile's major unpublished projects was his history of mathematics, based on a census of ancient and modern writers. Bainbridge resolved to do the same for astronomy, referring in letters to Ussher to 'my Astronomical History', under which he comprehended the history of technical chronology too.[59] His brief contribution to Hakewill may have derived from this project, which otherwise appears not to have materialized. This may be because Bainbridge then took the humanistic project one stage further than even Savile had been able to, and this started to engross his attentions, just as logarithms had captivated his colleague Briggs.

For Bainbridge realized that a great deal of astronomical knowledge was locked up in the texts of the medieval Islamic observers:[60]

It is a difficult thing which I undertake, but the great hopes I have in that happy *Arabia* to find most precious Stones . . . do overcome all difficulties.

To this end he acquired Arabic and Persian books, including a manuscript of the 15th-century Timurid Sultan astronomer Ulugh Beg, and started studying Persian – how, we are not sure – in order to read it.

Cl. V. IOHANNIS BAINBRIGII,
Aftronomiæ,
In celeberrimâ Academiâ Oxonienfi,
Profefforis Saviliani,
CANICVLARIA.

Unâ cum demonftratione
Ortus Sirii heliaci,
Pro parallelo inferioris Ægypti.

Auctore IOHANNE GRAVIO.

Quibus accefferunt,
Infigniorum aliquot Stellarum Lon-
gitudines, & *Latitudines,*
Ex Aftronomicis Obfervationibus
Vlug Beigi,
Tamerlani Magni nepotis.

OXONIÆ,
Excudebat HENRICUS HALL, Impenfis
THOMÆ ROBINSON, 1648.

(*Left*) Part of John Bainbridge's manuscript of Ulugh Beg.
(*Right*) Bainbridge's *Canicularia* was one of the earliest Oxford books to use Arabic type.

Bainbridge's work again remained largely unpublished, other than a portion that was edited and published by his successor John Greaves as *Canicularia*, a chronological work on the ancient Egyptian observations of the rising of Sirius, in which he corrected that greatest of European scholars, J. J. Scaliger, a feat recorded in the text later composed by John Greaves for Bainbridge's memorial in Merton College Chapel. Bainbridge thereby inaugurated an interest in Arabic and Persian sources that would attract subsequent Savilian professors, including John Greaves, John Wallis, and Edward Bernard. It also influenced Edmond Halley, who would teach himself Arabic in order to be able to edit Apollonius's *Conics*. Halley's edition of 1710 remains one of the great editorial achievements in mathematics, as we shall see in Chapter 3.

The second Savilians

We close this chapter with a few remarks on the second Savilian professors, Peter Turner and John Greaves. Turner, who was Briggs's successor at Gresham College, also followed

Si cupias VISITOR Quis et Quantus hic iacet
Alibi quæras oportet dicere satis nequeo,
BRITANNIA tota Viri famam non capit.
Ne cætera tamen ignores, in rem tuam pauca hæc accipe.

IOHANNES BAINBRIDGIVS

Vir famæ integerrimæ et Doctrinæ incomparabilis,
Medicinæ Professor et Mathe[s]eos,
Morborum tam foris expugnator Novorum
Quem longæ indagator Syderum
Quem Primû Astronomiæ Professor a digniss Savilio Collegii
In Arithmetica Professur, quas Magnifice erexerit
Prudens Hominis et Librorum Æstimator elegit
SAVILIVS

Quem CANTABRIGIÆ educatum
ACADEMIA OXONIENSIS benigne fovit, ut Stand,
Defunctum Publice celebrit, ut fui venitq, Ornamentum
Qui scientiarum sotibus contexit quæ pracica emendavit
TEMPORA.
In non levem Literarum iacturam
Immaturus Obijt
MDCXLIII.
Abi iam, cætera quære vel ab Exteris

John Bainbridge's memorial in the antechapel of Merton College.

him in his move from Gresham's professorship to that of Savile, with the qualification that Turner was already a Fellow of Merton College and had continued to reside chiefly in Oxford throughout.[61] We know that as a Savilian professor he lectured on magnetism, having read the *Philosophia Magnetica* (1629) of the Jesuit philosopher Niccolò Cabeo;[62] this again displays the easy overlap between the fields of astronomy and geometry, as the geometer was here presumably defending against Cabeo the cosmological arguments of the pioneering English magnetician William Gilbert, who had argued in 1600 in favour of the rotation of the Earth in his *De Magnete*.[63]

Peter Turner had a reputation as a linguist in the classical and oriental languages, particularly in Greek, and apparently translated from Greek into Latin the letters of the 5th-century ascetic Isidore of Pelusium – an obscure (but not an original) act. He also orchestrated transcriptions from Oxford manuscripts of the Ancient Greek writers on musical theory, Gaudentius, Alypius, and Aristides Quintilianus, for the scholar John Selden, and Ussher thanked him in print for a critical suggestion in Greek patristics. According to Anthony Wood,[64] Turner was

a most exact latinist and Grecian, was well skilled in the Hebrew and Arabic, was a thorough pac'd mathematician, was excellently well read in the fathers and councils, a most curious critic, a politician, statesman, and what not.

Turner did bequeath some Greek manuscripts to the Bodleian Library, but, suspiciously, many appear to have been Savilian manuscripts in the first place and were eventually returned to that library. An ardent Royalist in the Civil War, Turner has the dubious distinction of being the only Savilian professor to have been exchanged as a prisoner of war (in 1643), and his loyalty cost him his career. He died in poverty in London in 1652.[65]

But Turner's strictly mathematical achievements, if there were any, are now all but invisible to us. As far as we can see, Turner used his post chiefly to become an academic politician, Archbishop Laud's right-hand man in Oxford, and one of the authors of the Laudian Code, the University regulations that were in force until the mid-19th century. He also drew up the 'Caroline proctorial cycle', a table showing which Oxford colleges' turn it was to supply proctors, the officers who oversee academic discipline. It has been shrewdly conjectured that he was therefore responsible for the graphically similar *Encyclopaedia seu Orbis Literarum* of 1635, a pictorial chart that represented the Oxford curriculum, with the Sun at its centre and the planetary symbols around it – in other words, a Copernican schema.[66] But in truth we still know very little about him.

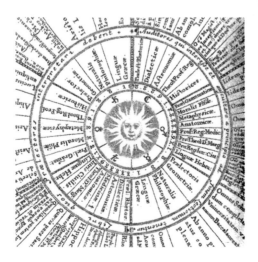

Peter Turner's *Encyclopaedia seu Orbis Literarum*.

It is quite a different picture with John Greaves.[67] Turner's counterpart in astronomy was one of the great Savilians, and again demonstrates that we should not be too hasty to separate out mathematical from astronomical interests. Greaves was one of four learned brothers who all achieved distinction in different fields, from Arabic to medicine. He had succeeded Turner as Gresham Professor of Geometry in London, only switching subject and post following Bainbridge's death in 1643, so that three out of the first four Savilian professors had transferred to Oxford from the geometry chair at Gresham College.

By that time, Greaves had already achieved extraordinary things, and as a scholar he was truly the heir of Bainbridge, whose mathematical orientalism he took to higher levels. With the encouragement of Archbishop Laud, Greaves set off for a tour of the Levant in 1637, in the company of Edward Pococke, the first Laudian Professor of Arabic. Greaves, who had already travelled extensively on the Continent, carried with him to the East a set of mathematical instruments that had been constructed by the Oxford maker Elias Allen. Picking up from where Bainbridge had stopped, he was adept enough in Persian by 1640 to write a grammar of it, which he printed nine years later in Oxford, and in the same decade he started work on a Persian dictionary at the request of Archbishop Ussher. While in Egypt, Greaves climbed into the Great Pyramid of Cheops in order to measure its inner chambers, as recounted in his most memorable book, the *Pyramidographia* of 1646, a work that once again balanced a humanistic concern for the evaluation of ancient accounts of the pyramids with a modern appreciation of actual

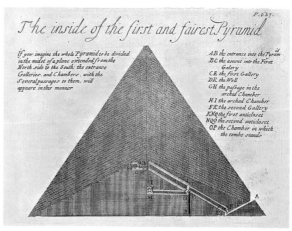

John Greaves (1602–52), the second Savilian Professor of Astronomy, and a diagram from his *Pyramidographia*.

hands-on measurement. Back at Merton College, he maintained the career link between astronomy and medicine, serving also as Linacre lecturer as Bainbridge had done before him.

Conclusions

The early Savilian chairs consolidated and improved mathematical instruction in Oxford at the University level; they certainly did not create it, and they accompanied college mechanisms for instruction also. Their real significance in pedagogical terms was their embedding of the senior tenured research professor in subjects other than the traditional higher disciplines of law, medicine, and theology. But to appreciate this we need to juxtapose the chairs with other developments, and especially the rise of the specialist college lecturer, the ancestor of the modern tutorial fellow: the oft-repeated claim that the modern tutorial system originated in the 19th century is quite false. Secondly, Henry Savile's own academic work and the spirit informing his foundations are best described as humanist, and this spirit informed the work of his early professors, from Briggs and Bainbridge's editing of Euclid and Ptolemy to Turner and Greaves's seeking out of ancient musical writers or medieval Persian astronomers. In retrospect, Savile's two chairs may symbolize the institutional establishment of university science in the West, an activity that has come to dominate (or even to tyrannize) modern higher education across the globe; but that is an anachronistic view, and at the time the Savilians were simply raising the standard of the humanities in Oxford.

Finally, despite his own taste for tough editorial problems rooted in Greek, Henry Savile was also aware that students needed practical know-how with instruments, and the ability to make and collate actual observations, in addition to their philological skills. The early professors did not neglect this aspect of their duties either. The balance is nicely captured in a letter from Bainbridge to Ussher in which he followed a long discussion of books he sought with the statement, 'I am very busy in the Fabrick of a large Instrument for Observations',[68] and, as we have seen, Bainbridge procured many new instruments for the Savilian collection. Indeed, Savile wrote in his statutes that his lecturers had to be instrumentalists, because in so doing they would be acting 'in imitation of Ptolemy and Copernicus' themselves.[69] We are now so used to histories of scientific revolution as narratives of disruption and innovation that we might pause to appreciate that, for Henry Savile, his professorships were about continuity – continuity with the past, and continuity into the future.

Portrait of John Wallis by Godfrey Kneller (c.1702). Samuel Pepys commissioned and paid for this portrait and presented it to Oxford, where it now hangs in the Examination Schools.

CHAPTER 2

John Wallis

PHILIP BEELEY AND BENJAMIN WARDHAUGH

Reading, writing, and doing mathematics in turbulent times, John Wallis (1616–1703) became the Savilian Professor of Geometry in unpromising circumstances, but held that position for longer than any other. Taking seriously the founder's injunctions to study, edit, and publish the ancient mathematical texts, as well as to teach mathematics, he also enjoyed a long career as a robust and combative mathematical author. In this chapter we consider Wallis's achievements as a reader, writer, and shaper of mathematics in the early modern world.

Introduction

In the long history of the Savilian professorships at Oxford, John Wallis's tenure of the geometry chair is unique. Not only has he been the longest serving incumbent – in all, he occupied the position for fifty-four years, from 1649 to 1703 – but no other holder of the post arrived in the way that he did.

Following the purge by the Parliamentary Commissioners of members of the University of Oxford who were deemed to have been too loyal to the old regime, both of the Savilian chairs were vacant by the time that England was declared a republic in May 1649. Wallis, whose services to Parliament as unofficial codebreaker had been both exceptional and successful, and who had made no secret of his desire to embark upon an academic career (preferably in the mathematical sciences), was elected to a Savilian professorship

Philip Beeley and Benjamin Wardhaugh, *John Wallis*. In: *Oxford's Savilian Professors of Geometry*. Edited by Robin Wilson, Oxford University Press. © Oxford University Press (2022). DOI: 10.1093/oso/9780198869030.003.0002 by the Parliamentary Visitors on 21 June 1649. Allowed the liberty of choosing which chair he preferred, Wallis 'made choise of the Geometry professor the place out of which Dr Turner had been formerly ejected & the place consequently voyd'.[1]

It might seem to us today that, prior to his election as Savilian professor, Wallis had little mathematical experience and no public reputation as a mathematician, his previous appointments having been exclusively theological. As he admitted:[2]

In the year 1649 I removed to *Oxford*, being then *Publick Professor of Geometry*, of the Foundation of *Sr. Henry Savile*. And *Mathematicks* which had before been a pleasing Diversion, was now to be my serious Study.

Although he was clearly mathematically talented, questions of political expediency were of overriding importance. Retrospectively, his election to the geometry chair was a remarkable stroke of good fortune, both for mathematics and for Oxford, as will become clear.

Born in 1616, during the reign of King James I, John Wallis experienced the Civil Wars, the Commonwealth, and the Protectorate. Following the Restoration of the monarchy in 1660, he became a leading member of the newly established Royal Society, while continuing to fulfil his professorial duties in Oxford.

He also continued to serve government as the country's leading decipherer, becoming holder of the first such official post, shortly before his death in 1703 during the reign of Queen Anne. As Wallis explained, his survival strategy through the vicissitudes of political life in England was to adopt a course of moderation, but this should not be taken to imply quiescence. It did not, for example, prevent him from adding his name to a pamphlet delivered on 18 January 1649 by more than fifty London ministers of the cloth, denouncing the purging of Parliament by the army and the trial and execution of King Charles I.[3] Nor did it prevent him from becoming embroiled in innumerable scientific disputes – most notably with Thomas Hobbes and with contemporary French mathematicians such as Gilles Personne de Roberval, Blaise Pascal, and Pierre Fermat. As so often with Wallis, his claim to moderation is largely true, but it pays for us to dig deeper.

Early years

We begin with John Wallis's studies at the University of Cambridge in the 1630s. As an undergraduate he attended Emmanuel College, a well-known centre of Puritanism at

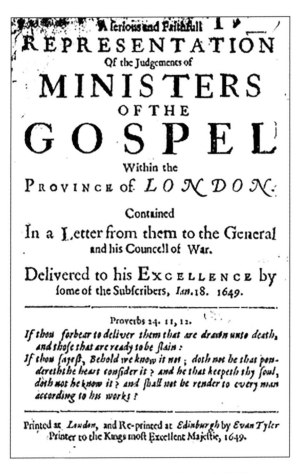

A ſerious and Faithfull I

REPRESENTATION

Of the Judgements of

MINISTERS

OF THE

GOSPEL

Within the

PROVINCE of *LONDON:*

Contained

In a Letter from them to the General
and his Councell of War.

Delivered to his EXCELLENCE by
ſome of the Subſcribers, *Ian.*18. 1649.

Proverbs 24. 11, 12.

*If thou forbear to deliver them that are drawn unto death,
and thoſe that are ready to be ſlain :
If thou ſayeſt, Behold we know it not ; doth not he that pon-
dereth the heart conſider it ? and he that keepeth thy ſoul,
doth not he know it ? and ſhall not be render to every man
according to his works ?*

Printed at *London,* and Re-printed at *Edinburgh* by *Evan Tyler*
Printer to the Kings moſt Excellent Majeſtie, 1649.

The 1649 pamphlet to which John Wallis appended his signature.

the time, where his tutors included Benjamin Whichcote and Thomas Horton, and his
student contemporaries included Ralph Cudworth, Jeremiah Horrocks, and John Wor-
thington. Wallis pursued the standard undergraduate curriculum, but demonstrated an
interest in the new philosophy by attending lectures by Francis Glisson on 'speculative
physics' and anatomy. Later, he would claim that the mathematical sciences were all but
completely neglected in England's two universities at that time: the following passage in
his autobiography has often been quoted:[4]

For Mathematicks, (at that time, with us) were scarce looked upon as Accademical studies, but rather Mechanical; as the business of Traders, Merchants, Seamen, Carpenters,
Surveyors of Lands, or the like; and perhaps some Almanak-makers in London.

This was clearly an overstatement on his part, for he also recalled having engaged while at Cambridge in 'Astronomy and Geography (as parts of Natural Philosophy) and . . . other parts of the Mathematicks'.

By this time, the two Savilian chairs already existed in Oxford, and we know that most of the colleges at the two universities offered instruction in mathematics, albeit to varying levels. At Cambridge, Wallis excelled in logic; two logical theses that he defended publicly were later appended to his *Institutio Logicae* (Foundation of Logic) of 1687, which became a standard text in that discipline and went through five editions up to 1729.[5]

Unable to gain a fellowship at Emmanuel College, Wallis left Cambridge in 1640, having graduated with a BA degree in 1636/7 and an MA degree shortly before he departed. In the same year he entered holy orders, ordained by Walter Curll, Bishop of Winchester and a close supporter of Archbishop Laud. He subsequently spent formative years in private chaplaincy, first in the Puritan household of the Darley family in Yorkshire. While there, he produced his first significant publication, a philosophical tract entitled *Truth Tried* (1643), which was conceived as a response to Robert Greville's widely circulated and discussed *Nature of Truth* of 1640. Much of this work he rejected, yet he subscribed to Greville's admiration for the pansophic writings of Jan Amos Comenius.

Thereafter, Wallis served as chaplain to Lady Mary Vere, an influential Puritan widow who resided partly in London, partly in the county of Essex. Through her, he was able to establish useful contacts in Parliamentary circles, but the most important of these came about by accident in 1642, when his innate skill in codebreaking was discovered. A visiting Parliamentary army chaplain challenged him to try his hand at an intercepted Royalist cipher. Wallis's unexpected success led to his becoming a prize asset to the Parliamentary cause during the Civil Wars, and subsequently to the Cromwellian government – and, as we have seen, it ultimately secured his professional future at Oxford.[6]

A more immediate result of the favour in which Wallis now found himself was his being presented with a sequestrated living in London and being made secretary or scribe to the Westminster Assembly of Divines, charged with reforming the Church of England. Alongside carrying out his ecclesiastical duties to the satisfaction of those around him, he used his presence in the metropolis to pursue further his interests in the new philosophy. In particular, he began attending regular scientific meetings emanating from Gresham College with like-minded men such as John Wilkins, Jonathan Goddard, Theodore Haak, Samuel Foster, and his former teacher Francis Glisson. Topics in the mathematical sciences, physiology, and all branches of natural philosophy were discussed at these

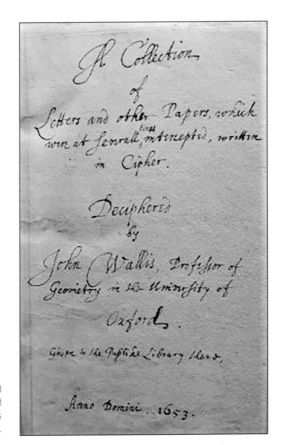

The title page of a collection of deciphered letters, many of them connected with the Presbyterian Conspiracy, which Wallis deposited in the Bodleian Library in 1653.

meetings, and occasional experiments were carried out. Wallis would always consider these London meetings, which were probably initiated by Haak in 1645 and which clearly made a great impression on him, to have been the true origin of the Royal Society, founded officially fifteen years later.[7]

Wallis as Savilian Professor

About a year before his election to the Savilian professorship, Wallis had begun to occupy himself seriously with the study of mathematics. We know that he worked his way systematically through William Oughtred's widely read *Clavis Mathematicae* (The Key to Mathematics), first printed in 1631 and significantly expanded in subsequent editions,[8] and developed an early interest in algebraic equations. In this sense largely self-taught, Wallis soon began to conduct investigations of his own on the nature of cubic equations,

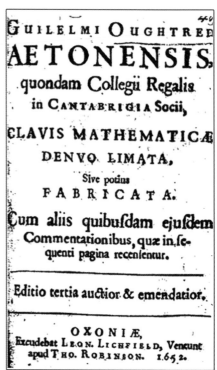

GUILELMI OUGHTRED
AETONENSIS,
quondam Collegii Regalis
in CANTABRIGIA Socii,
CLAVIS MATHEMATICÆ
DENVO LIMATA,
Sive potius
FABRICATA.
Cum aliis quibuſdam ejuſdem
Commentationibus, quæ in ſe-
quenti pagina recenſentur.

Editio tertia auctior & emendatior.

OXONIÆ,
Excudebat LEON. LICHFIELD, Veneunt
apud THO. ROBINSON. 1652.

GULIELMUS OVGHTRED ANGLVS.
ex Academia Cantabrigiensi Ætat: 73. 1646.

(*Left*) William Oughtred (1575–1660).
(*Right*) This 1652 edition of Oughtred's *Clavis Mathematicae* received editorial input from John Wallis.

and happened independently upon Cardan's rules for solving them.[9] Indeed, he achieved such competence that John Smith, Platonist and University Lecturer in mathematics at Cambridge, sought his advice on difficult passages in Descartes's *Géométrie*, to which Wallis had turned soon after completing his study of Oughtred. Here again we see evidence of his early desire to pursue an academic career. Smith was a Fellow of Queens' College, and Wallis was also made a Fellow there in 1644, but had to resign his Fellowship in the following year after his marriage to Susanna Glyde.

Wallis prepared for his future calling in another way, too. Early in 1649 he initiated a scientific correspondence with Johannes Hevelius after the intelligencer and educational reformer Samuel Hartlib had lent him a copy of the Danzig astronomer's study of the Moon's surface, the *Selenographia*, which had been published two years earlier. Wallis had considerable contact with Hartlib around this time, and Jonathan Goddard, a member of Hartlib's circle who knew Wallis through the scientific meetings that they attended together in London, had drawn his attention to the importance of Hevelius's work. Goddard would soon follow Wallis to Oxford, where he was appointed Warden of Henry

Savile's old college, Merton. Interestingly, Wallis's first letter to Hevelius, written just over a month before his election to the Savilian professorship, already contains strong pointers in that direction. Taking note of the enormous labour required in the calculation of astronomical tables, he offered to assist Hevelius. At the same time, he conceded that there were competing pressures to carry out the theological tasks for which he was paid. He would have left Hevelius and others in no doubt about which he would rather be doing.[10]

Buoyed by such strong motivations, Wallis's progress in mathematics was stupendous. Within the space of a few years he produced a series of significant mathematical works, as we now discover.

De Sectionibus Conicis Tractatus (Treatise on Conic Sections)

In this work of 1655 John Wallis treated the curves analytically, defining the parabola, the ellipse, and the hyperbola by means of algebraic equations, rather than as slices of a cone.[11] Here he spoke explicitly about treating an old topic by a new method – indeed, about liberating conics from a consideration of the geometrical object. In the second half of this book, he proceeded to find equations of tangents and other properties of the conic sections.

But his prime concern at this time was already with questions of quadrature – that is, finding areas. Indeed, in the very first proposition of his tract he introduced Cavalieri's method of indivisibles as a means of determining the areas enclosed by curves – or more accurately, the ratios of these areas to those of inscribed or circumscribed rectangles, the former being conceived as of infinitely small altitude. In fact, along with a number of his contemporaries, Wallis had not drawn this method from reading Cavalieri, but instead from the more widely distributed interpretation of Cavalieri's method by Evangelista Torricelli: we know that in the early 1650s Wallis systematically worked his way through the 1644 edition of Torricelli's *Opera Geometrica*.[12]

A notable feature of Wallis's book on conic sections is his introduction of new notation. Here we find the first appearances of the symbol ∞ for 'infinity' and the symbol $\geq$ for 'greater than or equal to'.

Arithmetica Infinitorum (Arithmetic of Infinities)

Wallis sought to proceed further with quadratures in another work that appeared shortly thereafter, his *Arithmetica Infinitorum* of 1656;[13] this would become his most widely read mathematical publication and constituted an important stage in the emerging field of

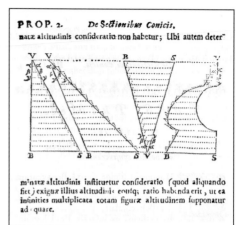

Two extracts from *De Sectionibus Conicis Tractatus* of 1655, exhibiting the symbol for infinity and Wallis's use of Cavalieri's method.

analysis. Newton studied it in depth and profitably, but Leibniz came to it late during his sojourn in Paris from 1672 to 1676 after he had already developed the foundations of his infinitesimal calculus.

Instead of using the algebraic techniques that he had developed in *De Sectionibus Conicis*, Wallis now adopted an arithmetical approach to the method of indivisibles. This was where he was perhaps most innovative, for Torricelli had proceeded geometrically, supposing that a plane figure was made up from an infinite collection of lines deemed to compose it, and in similar fashion that a solid was made up from an infinite collection of planes or surfaces.

Wallis saw how the necessary summations could be carried out arithmetically, by finding the sums of increasing sequences of terms such as an arithmetical progression (in the case of a triangle), or of squares or square roots (when determining the area of a parabola); Wallis's sequences began at 0 and had a finite number of terms. Decisively, when dealing with different cases, he moved from dealing with a finite number of steps to infinitely many, corresponding to the supposed composition of areas or volumes by infinitely many indivisibles. Thus, his treatment of quadratures and cubatures came to depend on the summation of infinite sequences of arithmetic infinitesimals, thereby replacing Cavalieri's geometry of indivisibles by his own arithmetic of infinities.

Wallis sought a general formula for the area under a curve of the form $y = x^n$, already known when n is a positive integer, and he extended the result to fractional and negative exponents. Starting with sequences of simple powers, he achieved his goal of finding the area inside a circle by making increasingly sophisticated interpolations. To this end he constructed tables of powers – an approach that was clearly inspired by his work as a codebreaker where there had often been a question of recognizing numerical regularities and patterns – and he arrived at the following infinite fraction for $4/\pi$, which he denoted by $\square$; this is now known as *Wallis's formula*:

$$\frac{4}{\pi} = \frac{3 \times 3 \times 5 \times 5 \times 7 \times 7 \times \text{etc.}}{2 \times 4 \times 4 \times 6 \times 6 \times 8 \times \text{etc.}}$$

While Wallis's derivation of this product was justly criticized, the result was confirmed by his pupil William Brouncker, who used a more rigorous method based on continued fractions.

As the Savilian Professor of Geometry, Wallis was required to deliver public lectures (open to all members of the University) on Euclid's *Elements*, the *Conics* of Apollonius, and all the known books of Archimedes. Alongside these foundational tasks, deeply rooted in the humanist tradition embodied by Henry Savile, there was to be instruction in other more accessible topics, such as theoretical and practical arithmetic, practical geometry or geodesy, mechanics, and music.

Only in respect of Euclid are we afforded any immediate insight into the form that Wallis gave to his lectures, for there survives a complete transcript in his hand, covering the years 1651 to 1652.[14] Interestingly, these were not introductory lectures, but focused instead on questions of interpretation based on the rich tradition of Euclidean

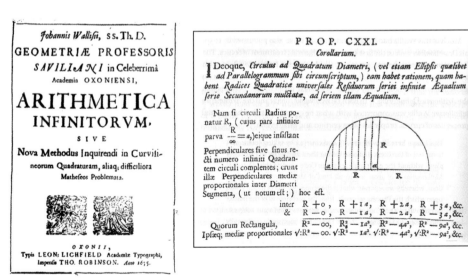

(*Left*) The title page of John Wallis's *Arithmetica Infinitorum*, published in 1656.
(*Right*) Proposition 121, in which Wallis investigates the area inside a circle.

PROP. CI XXXIX. *Theorema.*

H *Inc sequitur, quod* Si ex Tabellæ prop. 184. locis vacuis unus quilibet numero noto suppleatur, erunt & reliqui omnes cogniti.

Verbi gratiâ, si numerus hâc notâ □ designatus supponatur cognitus, reliqui omnes etiam cognoscentur; qui nempe eam habent ad illam rationem quæ hic subtus indigitatur.

Et (continuatâ ejusmodi operatione juxta Tabellæ leges) invenietur

$$\square \begin{cases} \text{minor quam } \dfrac{3 \times 3\times5\times5\times7\times7 \times9\times9 \times 11\times11\times13\times13}{2\times 4\times4\times6\times6\times8\times8\times10\times 10 \times 12\times12\times14} \times \sqrt{1\tfrac{1}{13}}. \\[2ex] \text{major quam } \dfrac{3\times3\times5\times5\times7\times7\times9\times9\times11\times11\times13\times13}{2\times4\times4\times6\times6\times8\times8\times10\times10\times12\times12\times14} \times \sqrt{1\tfrac{1}{14}}. \end{cases}$$

Et sic deinceps quousq; libet. Ita nempe ut fractionis Nu-

(*Left*) A table from the *Arithmetica Infinitorum*, involving the number $\square = 4/\pi$.
(*Right*) Wallis's upper and lower estimates for $\square = 4/\pi$.

commentary. Over a number of days, Wallis addressed philosophical topics such as the 'Quaestio de certitudine mathematicarum' (Question of the certainty of mathematics) or a debate over the nature of the angle of contact between two curves. It remains an open question as to how these lectures would have coordinated with college tuition, which was often based on plain editions of the first six books of the *Elements* or on such compendia as Pierre Gautruche's *Mathematicae Totius* (The Whole of Mathematics) of 1668.

Throughout the 17th century, mathematical teaching and learning remained patchy at Oxford and Cambridge. Although there were pockets of excellence that involved committed and able tutors and keen students who went on to become mathematicians of national or international reputation, the majority of those who studied at the English universities were indifferent to mathematics, had little contact with the subject, and acquired little of it by the time they graduated.

Wallis seems to have done little to change this. We search in vain for his students, or for reminiscences of his teaching; the manuscript text of his lectures on Euclid has survived, but we have nothing to indicate that he applied unusual energy to his teaching duties or that he undertook them with more success than his predecessors. But with another of Wallis's early publications we find ourselves in a slightly better position.

John Wallis's handwritten lectures on Euclid's *Elements*.

Mathesis Universalis, sive Opus Arithmeticum (General Mathematics, or Arithmetical Work)

John Wallis's *Mathesis Universalis* of 1657 is an elementary introduction to arithmetic.[15] Clearly the product of lectures that he delivered in his early years as Savilian professor, it is notable in that he traces the history of mathematical notation from antiquity to the present in both the Eastern and Western traditions. This reflects the important role that he ascribed to the history of mathematics in his writings, and also his approach to scientific problems in general:[16]

And (herein as in other studies) I made it my business to examine things to the bottom; and reduce effects to their first principles and original causes. Thereby the better to understand the true ground of what hath been delivered to us from the Antients, and to make further improvements of it.

Wallis dedicated his *Mathesis Universalis* to four of his most powerful Oxford contemporaries: Gerard Langbaine (Provost of The Queen's College), Henry Wilkinson (Canon of Christ Church and Lady Margaret Professor of Divinity), John Wilkins (Warden of Wadham College), and Jonathan Goddard (Warden of Merton College). Like Wallis and Goddard, Wilkins had been intruded by the Parliamentary Visitors, and was now the guiding spirit of Oxford's philosophical club which met regularly in his lodgings in Wadham College. In his dedication Wallis again declared his commitment to the

Savilian statutes in nurturing (and above all in promoting) mathematical studies – seeking, for example, to show how some propositions in Book II of Euclid's *Elements* can be effortlessly demonstrated in a purely algebraic manner.[17]

For Henry Savile there had been two blemishes in Euclid's *Elements*. One of these was the fifth postulate (otherwise referred to as the 'parallel postulate'), while the other was the fifth definition of Book VI.[18] Savile charged his successors with addressing these, and Wallis did not neglect to do so. On the evening of 11 July 1663 he delivered a lecture in Oxford on these topics, in which he argued persuasively that Euclid's fifth postulate could be derived from his other axioms. His argument was based on his deduction that if there were a geometry in which the first four postulates hold, but not the fifth, then any two geometrical objects with the same shape (such as two squares) must necessarily have the same size. Since this is clearly nonsensical, he claimed, no such geometry can exist.[19] This attempt by Wallis to derive the parallel postulate as a theorem was to become a notable episode in the history of 'non-Euclidean geometries', later to be investigated by Johann Heinrich Lambert and Girolamo Saccheri. It was not until the 19th century that such strange geometries were studied seriously as consistent mathematical structures in their own right.

Correspondence and controversy

Like no other English mathematician, Wallis grasped his new opportunities for promoting the growth of scientific knowledge through the free exchange of ideas. In his inaugural lecture as Savilian professor, delivered on 31 October 1649 in the Geometry lecture room of the Schools Quadrangle, he presented an overview of the mathematical tradition from the beginning of civilization, and suggested that it was much easier for him and his contemporaries to add to the discoveries of the Greeks than it had been for them to make their discoveries in the first place. His justification for this claim was the variety of means of communication that were now available, 'especially since it is now possible through the benefit of printing for ideas to be communicated across Europe which in those days were almost restricted to Athens'.[20] Alongside print, Wallis wrote many letters, soon adding scholars like Frans van Schooten, Christiaan Huygens, and Pierre Gassendi to his network of correspondents.

But there were disadvantages to such prominence. In coming to personify England's late emergence in discourse on the mathematical sciences, Wallis soon found himself drawn into controversies with mathematicians on the Continent – especially in France – who sought to test his capabilities and the originality of his ideas. Political and

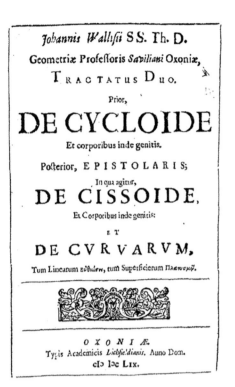

(*Left*) John Wallis's inaugural lecture as Savilian Professor of Geometry took place in October 1649.
(*Right*) His 1659 tract on the cycloid, in which he presented his solution to questions posed by Pascal.

military rivalries between the two nations also helped to fuel a quick succession of disputes with Roberval (over the rectification of the semi-cubical parabola, $ay^2 = x^3$), with Pascal (over the quadrature and cubature of the cycloid), and with Fermat (over questions in number theory).[21] Ultimately, such disputes would serve to dent his belief in the value of open scientific communication. They would also contribute to his unfavourable perception of Leibniz in the priority dispute over the discovery of the calculus, even though his own behaviour was certainly not always above reproach.

There were other disputes, too, but none came to overshadow Wallis's professional career so much as those with the philosopher Thomas Hobbes, which lasted for almost a quarter of a century. Wallis's Oxford colleague Seth Ward, the Savilian Professor of Astronomy, had accused Hobbes of plagiarizing the surviving manuscripts of Thomas Harriot's disciple Walter Warner, and had challenged him to provide evidence of his own mathematical discoveries. Wallis joined the assault on a thinker who was widely suspected in England's intellectual circles of espousing radical materialism, while Hobbes, for his part, had friends in high places, including the King, whom he had instructed in geometry as a young man in Parisian exile.

In his *Elenchus Geometriae Hobbianae* (Examination of Hobbesian Geometry) of 1655, Wallis set out to display to a wider scholarly audience the ineptitude of the solutions to ancient problems such as the quadrature of the circle and the duplication of the cube, which Hobbes had published shortly beforehand in his *De Corpore* (On Body). Hobbes responded in the following year with his acrimonious *Six Lessons to the Professors of the Mathematiques, one of Geometry, the other of Astronomy: In the Chaires set up by the Noble and Learned Sir Henry Savile, in the University of Oxford*. This work attacked both Wallis and Ward, to which Wallis in turn wrote a reply entitled *Due Correction for Mr Hobbes*. This intellectual war, in which belittlement of the opponent was increasingly used as a rhetorical tool, did more harm than credit to both men and continued on and off until Hobbes's death in 1679.[22]

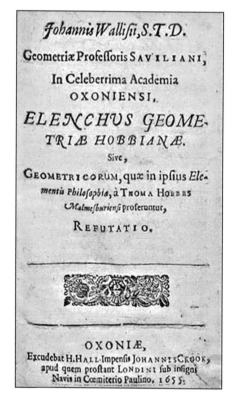 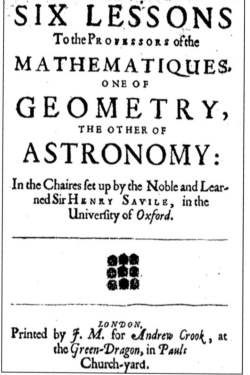

(*Left*) John Wallis's *Elenchus Geometriae Hobbianae*.
(*Right*) Thomas Hobbes's *Six Lessons to the Professors of the Mathematiques*.

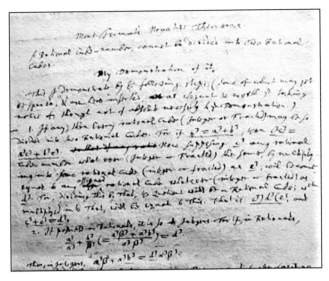

John Wallis's manuscript on Fermat's negative theorem: 'A rational cube cannot be divided into two rational cubes'.

John Wallis was always keen to make his mark beyond Britain's shores, and his work as a linguist was conceived directly with this aim in view. In 1653 he published his *Grammatica Linguae Anglicanae* (Grammar of the English Language), a largely traditional grammar with minor innovations, produced with the core purpose of providing an easy introduction to the English language for foreigners, at a time when the language was scarcely read abroad. He prefixed this work with what he considered to be a universal treatise on speech, *De Loquela*, which later served him as a model for the practical instruction of deaf mute individuals. But here again Wallis soon found himself in conflict, this time with the English clergyman and music theorist William Holder, who felt that his own efforts at such instruction had been disregarded. What made this dispute all the more difficult was that the two protagonists were both highly respected and scientifically active members of the Royal Society. It therefore raised questions about the very principles for which that institution stood.[23]

As revealed earlier through his work on logic, Wallis was a master of reasoned argument who repeatedly put his evident skills in dispute and refutation to use in his academic endeavours. He also evidently considered it important, and proper for him as Savilian professor, to play a part in policing the borders and the content of mathematics.

One telling example is the case of the Danish scholar Marcus Meibom, who in the early 1650s had prepared Greek and Latin editions of the entire corpus of Greek musical texts (barring three added later by Wallis); he was then in his 20s, a precocious, careful, and by most standards, brilliant scholar. In the wake of this musical editing project he took it upon himself to write a book-length account of mathematical ratios, an active subject of research that was of much relevance to the study of music, which often involved the elaborate consideration of the ratios of lengths of sounding strings. His book on ratios took the form of a dialogue whose participants included Euclid himself; Meibom's main point was that the proper manipulation and combination of ratios was being carried out wrongly by some of his contemporaries.

The merit of Meibom's arguments is moot, although they were not obviously absurd and his position concerning ratios was defensible. But his book, which appeared in 1655, raised the wrath of several mathematicians, including that of Wallis who proceeded to write *Adversus Marci Meibomii, De Proportionibus Dialogum Tractatus Elencticus* (Didactic Treatise against Marcus Meibom's Dialogue on Ratios). It was in print by 1657 and could scarcely have been more offensive about the book's alleged absurdities – at one point, Wallis said he initially thought Meibom was joking.[24]

Johannis Wallisii SS. Th. D.
Geometriæ Professoris *Saviliani,*
In Celeberrima Academia
OXONIENSI;

ADVERSUS
MARCI MEIBOMII,
De Proportionibus Dialogum,
Tractatus Elencticus.

OXONII,
Typis *Leonard. Lichfield* Academiæ Typographi.
Impensis *Tho. Robinson.* 1657.

John Wallis's *Adversus Marci Meibomii, De Proportionibus Dialogum Tractatus Elencticus* of 1657.

There is an air of a storm in a teacup about Meibom's book on ratios and the four responses that it elicited, but such things mattered in an era before peer review or printed book notices. Meibom never published on mathematics again; indeed, Wallis's disapproval in particular seems to have ended his attempt to enter the international community of publishing mathematicians. As with his more celebrated dispute with Hobbes, Wallis gave the strong impression that he considered it his role both to defend and to define the discipline of mathematics, pronouncing on who was allowed to participate and who was not. No one seems to have disputed his right to exercise that judgement, even though some would disagree with the judgements themselves.

Wallis and the Royal Society

Wallis relied on correspondence to keep abreast of developments in the scientific world, and to play an active part in discussions that were taking place at a considerable distance away in London. Henry Oldenburg, the Royal Society's corresponding secretary, was a decisive figure in this respect, communicating Wallis's frequent letters on mathematical and natural philosophical topics to members attending the weekly meetings and acting as an intermediary in subsequent exchanges. In return, Oldenburg supplied the Savilian professor regularly with the latest news, gleaned from his own extensive correspondence, while also occasionally sending him materials for perusal and assessment.[25] Alongside John Pell and John Collins, Wallis served as Oldenburg's foremost adviser on mathematical topics.

Despite being based in Oxford, Wallis was therefore able to become one of the most active members of the Royal Society, right through to the end of his life, and on more than one occasion he was called upon to help revive the institution when periods of non-participation of members threatened its collapse.[26] Some seventy articles or book reviews published in the Society's *Philosophical Transactions* are either under his name or can be reliably ascribed to him as author. Most are on mathematical topics, but taken as a whole they reflect the full breadth of his scientific interests: thus, alongside papers on the method of tangents or the true division of the meridians in a sea-chart, we find observational accounts of the tides in Kent (the county from which he hailed) or on the suspension of quicksilver in the Torricellian tube. He delivered numerous learned contributions that sought to explain natural phenomena based on experimental observation, including an important discourse on gravity, published in 1674.[27]

Wallis also contributed to a Royal Society debate on the laws of motion in 1668–69, in which his 'summary account of the general laws of motion' appeared alongside

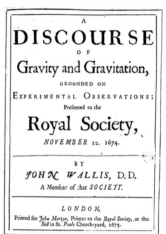

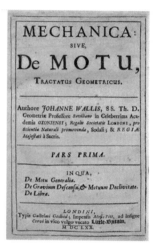

(*Left*) Wallis's *A Discourse of Gravity and Gravitation, Grounded on Experimental Observations*, presented to the Royal Society on 12 November 1674.
(*Centre*) His 'A summary account of the general laws of motion'.
(*Right*) His *Mechanica* of 1670–71.

those of Christopher Wren and (somewhat later) of Christiaan Huygens. Wallis derived these laws while he was working on his three-part *Mechanica: sive, De Motu, Tractatus Geometricus* (Mechanics: or a Geometrical Treatise on Motion) of 1670–71, published under the auspices of the Royal Society. This was a milestone in the mathematization of mechanics, containing a large section on the determination of centres of gravity. Moreover, it attracted the attention of Leibniz, who studied it in detail while in Paris, and who concurred with Wallis on his principle of the proportionality between effects and their sufficient cause, citing this as representing the gateway from mathematics to physics.[28]

Wallis had earlier been called upon by the Royal Society to review Leibniz's youthful tract *Hypothesis Physica Nova* (New Physical Hypothesis) of 1671. His enthusiastic account, in which he drew attention to considerable areas of agreement with his own ideas, contributed decisively to Leibniz's election to the Royal Society in 1673. Significantly, when discussing the causes of gravity and elasticity, for which Leibniz had proposed mechanistic explanations, Wallis took recourse to fundamental ideas on the nature of scientific truth, which again reflected his self-ascribed moderation:[29]

I am not one to rush in as an umpire where there is so great diversity of opinion. The question must be left to time and the arguments of the learned on either side. Indeed, almost the same thing happens with the swings of pendulums; after many oscillations on either side, at last they come to rest in the perpendicular.

We have seen this with the Copernican hypothesis which, though known to the ancients, lay buried so long that it was regarded as new; and although it had the strongest possible support it did not at once prevail but was attacked by different persons in different ways and bitterly disputed, until in the end through the ascendancy of reason over authority it was so universally acknowledged that virtually no one with any knowledge of the matter has any doubt about it, except those swayed by the Cardinals' decree.

Historical studies and classical editions

Moderation was less in evidence in Wallis's *Treatise of Algebra, both Historical and Practical* of 1685, comprising one hundred short chapters that ranged over algebra and its history.[30] As a history of algebra, the book was the first of its kind and included useful and informative discussions on methods of quadrature and on infinite series, as well as on the developing concept of mathematical proof. However, his *Treatise* was heavily biased toward English contributions to algebra, especially from Harriot and Pell, and in it Wallis set out, not for the first time, the unsubstantiated and false claim that Descartes had plagiarized the former in his *Géométrie*.[31]

One topic of interest in the *Treatise* was Wallis's geometrical construction for the square root of a quantity bc, when b and c are both positive numbers.[32] His construction involved drawing a circle with diameter AC of length $b + c$ and constructing a perpendicular from the point at distance b from A; the length of this perpendicular is then the required square root of bc. Wallis subsequently attempted to modify his process (not entirely satisfactorily) so as to construct the square root of bc when b and c have opposite signs, thereby hinting at the idea of an imaginary number.

Original mathematical publications were only part of Wallis's output in the discipline. In the mid-1660s he had been tasked by the Royal Society with collecting and selecting for publication the letters and papers of the Liverpool astronomer Jeremiah Horrocks, his prematurely deceased contemporary at Cambridge; this was his first foray into mathematical editing, but it almost failed for lack of money.[33] Horrocks's astronomical observations and ideas were rightly judged important enough to devote real effort to their presentation in print, and it was Wallis who undertook much of the work of arranging the texts for publication, also receiving assistance from John Collins and the astronomer John Flamsteed. Eventually, his edition of the posthumous works appeared in 1673 and included short pieces by William Crabtree and Flamsteed.

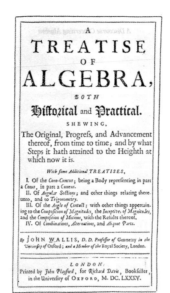

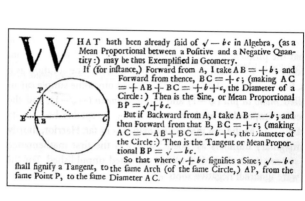

(*Left*) John Wallis's *A Treatise of Algebra, both Historical and Practical* of 1685.
(*Right*) His construction of a square root, from *A Treatise of Algebra*.

In 1676 Wallis brought out an edition of Archimedes' 'Sand-reckoner' (*Arenarius*) and 'Measurement of the Circle' (*Dimensio Circuli*), together with Eutocius's commentary on the latter. Later, in 1688, there followed critical editions of other Greek texts, including Aristarchus's work on the size and distance of the Sun and the Moon, and a previously unknown fragment from Book 2 of Pappus's *Collection* that set new standards in English textual scholarship.[34]

Wallis was also involved with a long-running project in Oxford to produce a new edition of the works of Euclid. His colleague in the astronomical chair, Edward Bernard, assembled to this end a vast quantity of textual and annotation material in a somewhat chaotic manner, and sought unsuccessfully to find a publisher. When the project was restarted with Bernard's successor David Gregory as named editor, Wallis was tasked with making sure that this time everything went to plan, the magnificent edition eventually being published in 1703.[35]

In addition to all this, Wallis developed and maintained throughout his academic life an interest in the mathematical theory of music. This was another subject on which there were ancient Greek texts of great interest, most of which had been edited in their original language and translated into Latin by the middle of the 17th century. Music, or 'harmonics', was one of the seven liberal arts and remained notionally part of university arts curricula into the 17th century, being mentioned by name in the

Savilian statutes. Wallis evidently considered this subject part of his purview and, as well as writing letters and papers on such topics as the classical divisions of the scale, he also prepared editions of the three Greek texts that had not been included in Meibom's mid-17th century edition.[36] Thus in 1682 he brought out the three books of Claudius Ptolemy's *Harmonics* in Greek and Latin, with an extensive appended commentary.

Wallis reprinted this work in the final volume of his monumental *Opera Mathematica* (Mathematical Works), whose three volumes were published by Oxford's University Press at the Sheldonian Theatre between 1693 and 1699. Also included were an edition of Porphyry's musical commentary and an edition of the *Harmonics* of Manuel Bryenne. These were sophisticated and learned scholarly editions, bringing together the readings of eleven different manuscripts for the Ptolemy text, and reporting this evidence with a precision that was innovative in its day; indeed, they remained the best editions of the three texts until the 20th century.

Cryptographer

In earlier years, Wallis and Leibniz had corresponded through Oldenburg as intermediary, but in the 1690s, when they began to exchange letters directly, two themes were in the foreground: cryptography, and the priority dispute between Newton and Leibniz over the discovery of the calculus. Both were motivated by Leibniz's interests at the time, but they had also come to dominate Wallis's intellectual activity in his final years.

Wallis had been engaged by successive governments as a codebreaker, and this was doubtless a substantial reason for his remaining in post as Savilian Professor of Geometry throughout the Restoration and beyond. The regicide Thomas Scot, in an account of his own intelligencing activity provided shortly before his execution, vouched for Wallis's skill as a cryptographer, indicating the disinterested nature of his work and emphasizing his value to any future government:[37]

The Kings transactions with the Presbyterian Ministers . . . were made knowne to mee . . . very much more by letters intercepted which commonly were every word & syllable in Cypher, and deciphered by a learned gentleman incomparably able that way, Doctor Wallis of Oxford (who never concerned himself in the matter, but only in the art & ingenuity); it is a jewell for a Princes use & service in that kind.

John Wallis's *Opera Mathematica*, published at the Sheldonian Theatre, attest to a remarkable career of promoting mathematical studies and publishing at Oxford. This third volume contained his contributions to the Oxford ancient texts project: his editions of works by Ptolemy, Archimedes, Aristarchus, Pappus, and others, and the musical items of Porphyry and Bryenne.

The accession of William III to the throne in 1689 brought with it an increased workload for supplying political and military intelligence, particularly for the allied German state of Brandenburg–Prussia. Leibniz, who openly described Wallis as Europe's greatest living cryptographer, sought unsuccessfully to persuade the English mathematician to divulge his methods as a way of enriching his philosophical programme of the 'art of discovery'. At the same time, Wallis used the publication of his *Opera Mathematica* in the 1690s not only to present examples of his cryptographic skill, but also to set out a rather tendentious factual record of what Newton had revealed to Leibniz of his method of fluxions in 1676. This, more than anything else, led to an irreparable breakdown in trust between the two men.[38]

Survivor

Wallis's professional career was a model of survival in turbulent times, an intricate construct of duties and responsibilities to the State and to his academic institution. He combined his wide-ranging scholarly endeavours with the assiduous defence of Oxford's ancient privileges through the office of *Custos Archivorum* (Keeper of the Archives), to

John Wallis's, 'Reasons Shewing the Consistency of the Place of Custos Archivorum with that of a Savilian Professor'.

which he had been elected through an act of strategic, but not wholly legitimate, foresight in 1658.[39]

Quite simply, Wallis soon made himself indispensable to the University, both academically and institutionally. He was one of the Royal Society's most active members, and was always at hand to provide support in times of crisis. He was an important adviser to churchmen on theological issues, and for historical and religious reasons fiercely opposed the introduction of the Gregorian calendar in England. Over the course of his lifetime he published numerous theological discourses and sermons, most of which he had delivered in the University Church of St Mary.[40]

Finally, he had been a reliable servant of government, ultimately becoming the country's first official decipherer. Wallis himself saw the key to his professional survival as having been his ability to avoid the extremes of contemporary politics, and at the end of his autobiography, *Pro Vita Sua*, penned for his friend Thomas Smith on 29 January 1696/7 (8 February 1697), he summed up his life and times in the following words:[41]

It hath been my Lot to live in a time, wherein have been many and great Changes and Alterations. It hath been my endeavour all along, to act by moderate Principles, between the Extremities on either hand, in a moderate compliance with the Powers in being, in those places, where it hath been my Lot to live, without the fierce and violent animosities usual in

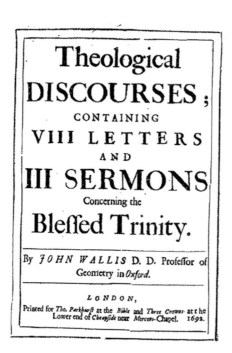

John Wallis's *Theological Discourses; containing VIII Letters and III Sermons Concerning the Blessed Trinity.*

Letter from John Wallis to Thomas Smith on 29 January 1696/7.

such Cases, against all, that did not act just as I did, knowing that there were many worthy Persons engaged on either side. And willing whatever side was upmost, to promote (as I was able) any good design for the true Interest of Religion, of Learning, and the publick good; and ready to do good Offices, as there was Opportunity; And, if things could not be just, as I could wish, to make the best of what is: And hereby, (thro' Gods gracious Providence) have been able to live easy, and useful, though not Great.

Arißippus Philosophus Socraticus, naufragio cum ejectus ad Rhodiensium litus animadvertisset Geometrica schemata descripta, exclamavisse ad comites ita dicitur, Bene speremus, Hominum enim vestigia video.
Vitruv. Architect. lib.6.Praef.

The frontispiece of Edmond Halley's 1710 edition of Apollonius's *Conics* depicts the story of Aristippus, introduced to Oxford audiences 140 years earlier by Henry Savile and adapted for Oxford publications over the next two centuries. It relates how the Socratic philosopher, shipwrecked on the island of Rhodes, was assured of the local inhabitants' civilized nature on discovering mathematical drawings in the sand.

A century of astronomers: from Halley to Rigaud

ALLAN CHAPMAN AND CHRISTOPHER HOLLINGS

Throughout the second half of the 17th century, John Wallis had successfully used his position as Savilian Professor of Geometry to promote pure mathematical research in Oxford. His successor, Edmond Halley (1656–1742), had different priorities. Now chiefly remembered as an astronomer, Halley took up the geometry chair following Wallis's death, having previously failed to secure the chair of astronomy when it became vacant in 1691. Nevertheless, although he did produce some mathematical writings (most notably, his edition of Apollonius's mathematical works), his main interests remained in astronomy. In this respect, he is characteristic of several of his 18th-century successors in Savile's chair of geometry, two of whom eventually moved sideways to take up the chair of astronomy.

Moreover, during the 18th century we find other occupants of the astronomy chair who are remembered for works in mathematics. Indeed, at this time, in the words of John Fauvel, 'the two Savilian Chairs seem to have been treated as essentially interchangeable in terms of qualification or inclination'.[1] In this chapter, we outline the mathematical and astronomical works of six successive Savilian professors of geometry between 1704 and 1827: Edmond Halley, Nathaniel Bliss, Joseph Betts, John Smith, Abraham Robertson, and Stephen Rigaud, setting them against the background of mathematics teaching in

Allan Chapman and Christopher Hollings, A century of astronomers: from Halley to Rigaud. In: *Oxford's Savilian Professors of Geometry.*
Edited by Robin Wilson, Oxford University Press. © Oxford University Press (2022). DOI: 10.1093/oso/9780198869030.003.0003

18th-century Oxford. In some instances, there is little to say – but this is certainly not the case with our first subject, Edmond Halley.

Edmond Halley

Halley became perhaps the most famous, and certainly the most widely published, English astronomer of his day – and his fame has lasted until the present time. His impact was felt not only in Oxford, but also nationally and internationally, and he was the first of three Oxford men to serve consecutively as Astronomer Royal, the third of these being Halley's Savilian successor, Nathaniel Bliss.

At the time of Halley's birth in Middlesex in 1656, astronomy was advancing rapidly. In the half-century since Galileo's first telescopic discoveries, observational planetary astronomy had come into being and had effectively challenged traditional evidence for the geocentric cosmology. New observational instruments, such as the micrometer and the pendulum clock, had been developed and the first (albeit unsuccessful) attempts to measure the distances of the stars were made by Christiaan Huygens in Holland. In the Oxford of the 1660s, Christopher Wren, as Savilian Professor of Astronomy, developed an international reputation with his studies in dynamics that later provided important links between the works of Kepler and Newton. Wren was knighted in 1673 and moved to an even more illustrious career as an architect.

Upon entering The Queen's College in 1673 at the age of 17, Edmond Halley is said already to have had a good knowledge of mathematics and astronomy: the antiquary Anthony Wood later asserted that Halley[2]

not only excelled in every branch of classical learning, but was particularly taken notice of for the extraordinary advances he made at the same time in the Mathematicks. In so much, that he seems not only to have acquired almost a masterly skill in both plain and spherical Trigonometry, but to be well acquainted with the science of Navigation, and to have made great progress in Astronomy before he was removed to Oxford.

Halley's facility with Greek, Latin, and Hebrew, which would enable him to study ancient and medieval mathematical texts, was apparently also acquired prior to his arrival in Oxford. Moreover, at this young age, Halley was already making observations with the fine private collection of astronomical instruments provided by his London merchant father. These included telescopes for observing the surface details of the Moon and the planets, incorporating such recent technical innovations as telescopic sights and micrometers.

EDMVND. HALLEIVS LL.D.
GEOM. PROF. SAVIL. & R.S. SECRET.

A Royal Society portrait by Thomas Murray of 'young Halley' in c.1687, holding a diagram of his work on cubic and quartic equations. The inscription relating to his Savilian chair was added later.

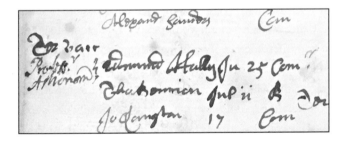

Edmond Halley's appearance in The Queen's College Entrance Book.

Astronomy was a current preoccupation in Oxford. In the year that Halley arrived, John Wallis published his edition of the papers of Jeremiah Horrocks, the innovative Lancashire observational astronomer who had died young in 1641 (see Chapter 2).[3] Within two years, Halley had made contact with John Flamsteed, recently appointed as the first Astronomer Royal, and began to collaborate with him, making observations at Oxford and travelling to assist him at Greenwich. In the Royal Society's *Philosophical Transactions* for 1675, Flamsteed remarked:[4]

Edmond Halley, a talented young man of Oxford, was present at these observations and assisted carefully with many of them.

Leaving Oxford without a degree in 1676, the 20-year-old Halley sailed to the island of St Helena with a set of new instruments to map and explore the skies of the southern hemisphere. It was Halley's work with Flamsteed, we may presume, that had awakened him to the opportunity of contributing to astronomical work in a spectacular way that was to secure him an international reputation by the age of 22. The foundation of the Royal Observatory at Greenwich in 1675 had recognized the official need to re-map the heavens at what everyone expected to be public expense. But the new observatories at Greenwich and Paris were concerned with making better catalogues of the long-familiar northern heavens. Halley's proposal to do the same for the southern skies was supported by several influential patrons, using both his Oxford and his family connections. Behind the letter from King Charles II that asked the East India Company to transport Halley and a friend to St Helena, there lay a web of influence – including such luminaries as Sir Joseph Williamson (the Secretary of State, and a Fellow of The Queen's College), Lord Brouncker (the President of the Royal Society and another former Oxford figure), and the eminent practical mathematician Sir Jonas Moore.

Although Halley was not the first European to map the southern skies – navigational charts that showed them had been commercially available for 150 years – he was the first to do so with the new level of accuracy made possible by telescopic measuring instruments. Using a large sextant equipped with telescopic sights, a quadrant, a pendulum clock, and micrometers, he obtained right ascension and declination angles to within about twenty arc seconds, greatly improving the several minutes of error in the parameters of existing charts. He also marshalled several miscellaneous groups of stars into new constellations, one of which he diplomatically christened 'Robur Carolinum' (Charles's Oak), a patriotic gesture that did not go unnoticed in the Palace of Whitehall.

In addition to mapping the southern skies, Halley was fascinated by the mysterious objects that they contained and which were not visible from European latitudes. These objects included nebulae and star clusters, classes of objects that would interest Halley for the rest of his long life. He discovered the star clusters of the constellation of Centaurus – perhaps the first in a long list of deep-space objects to be subsequently discovered in the southern hemisphere and of significance to the modern astrophysicist.

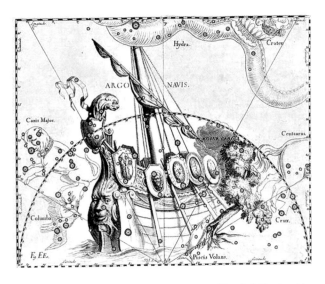

Halley's 'Robur Carolinum', as depicted in Johannes Hevelius's *Uranigraphia*.

The establishment of Halley's reputation

When his *Catalogus Stellarum Australium* was published, following his return home in 1678, Halley found that his reputation as an astronomer was made. The Royal Society elected him one of its youngest ever Fellows at the age of 22, he was acclaimed in Paris, and Charles II ordered Oxford University to grant Halley his MA degree by Royal *Mandamus*, rather than troubling him to be examined for it. This may well have been the first university degree to be conferred in explicit recognition of research achievement.

It also says a great deal about the young Halley's standing as a practical astronomer and his potential as a diplomat that the Royal Society chose to send him to Danzig in 1679 to adjudicate in an acrimonious dispute between Robert Hooke and the 68-year-old Johannes Hevelius. Hooke, who had done more than any other single astronomer to develop and advocate the use of telescopic sights for astronomical measurement, had severely censured Hevelius for failing to use them, and an unsavoury volume of abuse had arisen. That a young man aged just 23 should have been sent to comment upon the observing practices of the elder statesman of European astronomy is remarkable in itself, but the fact that he accomplished his task with so much success and tact speaks volumes for Halley's scientific and diplomatic skills.

Although Oxford University had launched Halley's astronomical career, he was to hold no formal status or appointment within the University for a further twenty-six years, although he remained on familiar terms with the place and was always regarded as an 'Oxford man'. Moreover, many areas of research that he initiated during these years were brought to maturity after his return to Oxford as Savilian Professor of Geometry in 1704. However, Halley was not one to be attracted to routine teaching, and at this stage of his career he would have found the University too narrow and provincial for his intellectual energies. It was through his involvement with the Royal Society, at the heart of metropolitan and international English science, that he was able to exercise those talents with which he was best endowed – research, administration, and diplomacy.

Halley's temperament and social skills also played a strong role in his success. He was fortunate in combining one of the most original scientific minds of his time with an astute social sense, an affable disposition, a strong constitution, a keen sense of humour, and financial shrewdness. Although Halley was the son of a prosperous London merchant, his private means do not appear to have been sufficient and he seems to have needed to earn his living. Indeed, his financial position presents the historian with something of a puzzle, for while he was capable of undertaking extraordinary acts of generous patronage, he frequently sought paid work and had the reputation for possessing a sharp eye for a profitable venture. He was an entrepreneur by instinct. Halley's marriage in 1682 to Mary Tooke, the daughter of an Exchequer auditor, was also singularly successful. It probably brought a dowry, and the marriage lasted for fifty-four years – an astonishingly long time, and against all the odds in an age when a person was lucky to live beyond 40.

Further evidence of Halley's diplomatic skills may be found in his apparently easy interactions with the notoriously temperamental Isaac Newton. Despite being vastly different in character, the two men were alike in having diverse and roving intellectual interests (though usually in different subjects), encompassing topics both mathematical and astronomical. Moreover, they shared a lifelong interest in the process by which physical motion in the heavens could be expressed in precise geometrical terms, thereby linking external observed phenomena in nature with the predictive and analytical capacities of systematic thought.

Halley's greatest contribution in this connection was probably his urging of Newton to write the *Principia Mathematica*, and indeed his payment of the costs of its 1687 publication by the Royal Society. Halley's closeness to Newton led to his being dubbed 'Newton's Creature' in the private papers of John Flamsteed, whose relations with Halley gradually deteriorated into acrimony, owing at least in part to the Astronomer Royal's jealousy of what he saw as Halley's easy success, and the latter's neglect of his former mentor.[5] Flamsteed certainly disapproved of Halley's levity (a view shared by Newton), as not becoming of a serious natural philosopher. Indeed, Flamsteed's negative feelings towards Halley were such that in 1691 he sought to ruin the latter's application for the Savilian chair of astronomy by informing Oxford that Halley would 'corrupt the youth of the University' with his scandalous ideas and conversation.[6] Despite positive testimonials from both the Royal Society and The Queen's College, who judged Halley 'to be every way most fit and accomplished' for the chair,[7] Flamsteed's influence won the day.

Halley's other interests

As we have seen, and will explore further below, Halley's major interests were astronomical. Indeed, we cannot neglect to mention his study, using Newtonian methods, of cometary orbits, leading to his establishment of their periodic nature, and his being memorialized in the name of Halley's comet. At a more earthly level, he also studied the variability of the Earth's magnetic field, and produced the first chart of that variation. Although his initial idea of applying geomagnetism to determine longitude at sea proved to be unworkable, his chart of magnetic variation is notable for its visual innovation of using lines to join points of equal value.

Alongside Halley's strictly scientific pursuits, we note also his interest in the history of astronomy. Many of the great astronomers of the 17th century prefixed the publication of their own achievements with an account of the history of the science up to their own time. Indeed, in Chapter 2 we saw a parallel pattern in the mathematical content of John Wallis's *Treatise of Algebra* of 1685. Riccioli, Hevelius, and others also gave credit to their predecessors, and especially to their near contemporaries such as Tycho Brahe, by way of setting the scene for their own work, while John Flamsteed's monumental *Historia Coelestis Britannica* of 1725 begins with a major history of astronomy.

All of these histories were intended to show a forward progress of achievement. Halley, however, displayed a different approach towards historical evidence, seeing it as a body of information that could be used to amplify modern understanding: in this respect he was following in a tradition of Oxford astronomy stemming from Sir Henry Savile and the first Savilian professors. Halley was an assiduous collector of astronomical data from previous centuries. Besides the famous historical studies upon which his cometary work depended, he collected evidence of ancient eclipses as a way of checking the new lunar theory based on Newtonian gravitation, as well as ancient inscriptions (such as those from Palmyra[8]) that recorded astronomical events.

But he was also aware of the great divide that separated the astronomers of his own century from earlier ones, as a series of recent inventions had transformed the quality and quantity of data almost beyond recognition. The telescope, pendulum clock, and micrometer allowed astronomers to see and measure things in the sky that had been

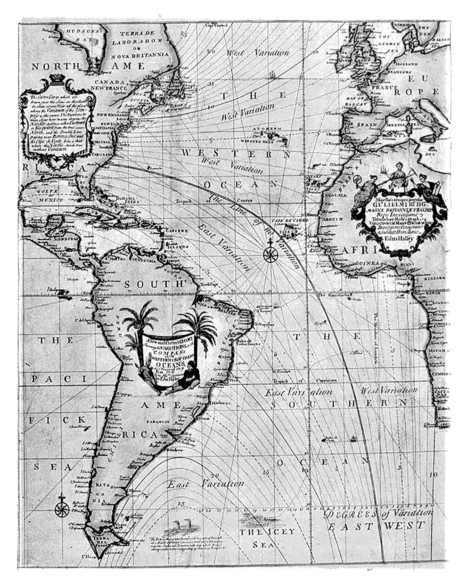

Halley's chart of magnetic variation over the Atlantic introduced the important idea of using lines to join points with the same value.

quite beyond the bounds of perception only a few decades before. In addition to his awareness of the significant observations made in the past, Halley also saw his own work and that of his contemporaries as forming terms of reference for future researchers.

Concerned as we are today with the public understanding of science, it is also of interest to note that Halley recognized a scientist's duty to educate the public. Two advantages that clearly qualified Halley for this role were his well-attested gift as a speaker and his clarity as a writer of prose. He relentlessly informed the readers of the *Philosophical Transactions* on a wide range of natural events, and also touched upon mathematical topics in a more-or-less accessible manner: a short piece of 1692, for example, discussed different types of infinite quantities in a largely non-technical manner, while another much-celebrated paper of 1693 contained an early example of the analysis of population data.[9] Halley also produced detailed articles on eclipses, meteors, meteorology, and other natural phenomena: in 1716, for instance, he published a paper on the strange brightness of Venus seen on 10 July of that year, in a direct attempt to explain a long-term characteristic of the planet's orbit currently being exploited by astrologers and scaremongers. For, as Halley emphasized:[10]

It may justly be reckoned one of the principal uses of mathematical sciences, that they are in many cases able to prevent the Superstition of the unskilful vulgar.

An awareness of the great cyclical patterns of nature, which he could substantiate from his historical and mathematical interests, gave him a capacity to educate the public in the new science.

The Royal Society's *Philosophical Transactions*, the only scientific magazine of its time, was read by a wide range of intellectually inclined people, and not just by active scientists. By writing for it, one obtained access to a broad range of intelligent readers. A wide penumbra of journalists, playwrights, and hacks further combed its pages for unusual material that could be recycled in newspapers and broadsheets. All the mad 'projectors' or scientists whom Jonathan Swift parodied in the Lagado Academy in *Gulliver's Travels* (1726), for instance, came from barely distorted reports of what appeared in the *Philosophical Transactions*. Although Halley might occasionally have opened himself up to satire, he was also able to get his ideas, and those of contemporary scientific thinking, directly into the public domain.

Halley as Savilian professor

In 1704, at the age of 48, Halley succeeded John Wallis as the Savilian Professor of Geometry – essentially his first proper academic post.[11] Few front-rank scientists had established their reputations in circumstances more removed from the security of the ivory tower than did Halley. That he was able to forge such a career path for himself in an age when there was no established 'career route' for an astronomer speaks volumes about his focus, determination, and ruthlessness. Indeed, he was always on the lookout for opportunities, prestige, and profits, a legacy perhaps of his mercantile background. Halley's scientific reputation had by now grown to such an extent that not even the Astronomer

The path of the total solar eclipse of 22 April 1715. Halley distributed this map throughout England to solicit observations of local eclipse conditions and reassure citizens that the eclipse was a natural event.

Royal could sabotage his application for the Savilian chair.[12] A letter of December 1703 from John Flamsteed reveals his irritation at the turn of events.[13]

Dr Wallis is dead – Mr Halley expects his place – who now talks, swears and drinks brandy like a sea captain.

Returning to Oxford after a quarter of a century, Halley set about securing his academic reputation within the University with a widely admired inaugural lecture:[14]

Mr Hally made his Inaugural Speech on Wednesday May 24, which very much pleased the Generality of the University. After some Complements to the University, he proceeded to the Original and Progress of Geometry, and gave an account of the most celebrated of the Ancient and Modern Geometricians. Of those of our English Nation he spoke in particular of Sir Henry Savil; but his greatest encomiums were upon Dr Wallis and Mr Newton, especially the latter, whom he styled his Numen etc.

One of Halley's first acts as Savilian professor was to publish his cometary researches with a historical introduction,[15] but over the next few years he justified the geometrical nature of his position by wisely adopting a suggestion from Henry Aldrich, Dean of Christ Church, that he take up a task left unfinished by his predecessor, John Wallis. He prepared what was to become the definitive edition of Apollonius's *Conics*, the comprehensive ancient text on the properties of conic sections. As part of the project, Halley also included a minor work of Apollonius, *De Sectione Rationis* (On the Cutting-off

IV. *Aſtronomiæ Cometicæ* Synopſis, *Autore* Edmundo Halleio *apud* Oxonienſes *Geometriæ Profeſſore* Saviliano, & Reg. Soc. S.

VEteres *Ægyptii* & *Chaldæi*, ſiqua fides *Diodoro Siculo*, longa obſervationum ſerie inſtructi, Cometarum ἐπιτολὰς prænuntiare valuerunt. Cum autem iiſdem artibus etiam Terræ-motus ac tempeſtates prævidiſſe dicantur, extra dubium eſt Aſtrologiæ potius calculo fatidico, quam Aſtronomicis motuum Theoriis eorum de his rebus ſcientiam referendam eſſe. Ac vix alia à Græcis utriuſque populi victoribus reperta eſt apud eos doctrina ; adeo ut eam , quam nunc eouſque provEximus Aſtronomiam , Græcis ipſis, præſertim magno *Hipparcho,* uti inventoribus, acceptam debeamus. Apud hos vero *Ariſtotelis* ſententia, qui Cometas nihil aliud eſſe voluit quam Vapores ſublunares vel etiam Meteora aerea, tantum effecit, ut hæc Aſtronomicæ ſcientiæ pars longe ſubtiliſſima, omnino neglecta manſerit, cum nemini operæ pretium viſum fuerit, vagas & incertas fluitantium in æthere vaporum ſemitas adnotare ſcriptiſque mandare ; unde factum ut ab illis nihil certi de motu Cometarum ad nos tranſmiſſum reperiatur.

Halley's 1705 publication of his cometary researches, with a historical introduction that maintained the Savilian tradition of grounding current scholarship in its historical roots.

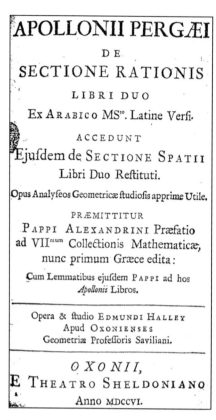

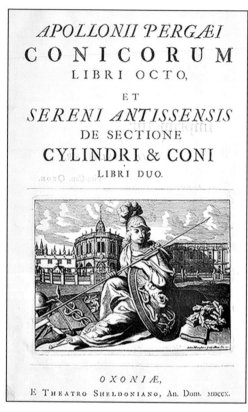

The first part of Halley's edition of the works of Apollonius, published in Oxford in 1706, and the completed work, published in 1710.

of a Ratio), which had survived in an Arabic manuscript, and a reconstruction of the fragmentary *De Sectio Spatii* (On the Cutting-off of an Area); these were published together in 1706.[16] Always an astute judge of people and situations, Halley must have realized that a scholarly edition of such a major classical geometer would silence those critics who accused him of being a gadfly and an opportunist, and secure him the respect of the academic establishment.

Halley worked on the *Conics* in conjunction with David Gregory, his Savilian co-Professor of Astronomy. Using Greek and Arabic manuscript sources, and guided by previous partial Latin translations, Gregory was to edit the earlier parts of the *Conics*, for which a Greek text was available, while Halley acquired a knowledge of Arabic in order to prepare the parts that survived only in that language. However, following Gregory's death in 1708, much of the work fell to Halley, including a reconstruction of the lost eighth book of the *Conics* from comments made by Pappus of Alexandria and an Arabic summary.[17]

PROP. XI. THEOR.

*In omni triangulo Sphærico, produǁo uno latere, angulus
exterior minor erit utrifque interioribus eidem oppofitis
fimul fumptis: & tres anguli trianguli fimul fumpti
majores erunt duobus reǁis.*

Sit A B Γ triangulum Sphæricum : dico quod angulus exterior,
arcubus B Γ, Γ Δ contentus, minor eft angulis A, B eidem oppofitis : quodque tres anguli trianguli A, B, Γ fimul fumpti excedunt duos angulos reǁos.

Fiat ad punǁum Γ fuper arcum Γ Δ angulus Δ Γ B æqualis angulo A, & producatur
A B ad occurfum ipfius A Γ in
punǁo Δ. Jam quoniam anguli Δ, Γ funt æquales, erunt
arcus Δ B, E Γ quoque æqua-
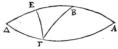
les ; & B E, E Γ fimul funt æquales arcui B Δ, ac proinde minores funt femicirculo : quare (*per præced.*) angulus exterior
Γ B A major eft interiore B Γ B, atque adeo angulus B Γ Δ exterior trianguli A B Γ, minor eft angulis Γ B A, B A Γ fimul fumptis.
Adjiciatur communis angulus B Γ A ; & duo anguli B Γ Δ, B Γ A
minores erunt angulis A, B, Γ. Sed duo anguli B Γ Δ, B Γ A funt æquales duobus reǁis : quare tres anguli A, B, Γ fimul fumpti
excedunt duos reǁos. Q. E. D.

Halley's Latin translation of Menelaus's proposition that the angles of a spherical triangle add up to more than 180°.

The work was eventually published in 1710 and clearly won the University's mark of approval,[18] for Halley was rewarded with an honorary Doctor of Civil Laws degree and a warm citation commending his public services from the University's Chancellor, the Duke of Ormonde. Halley's edition of Apollonius's works was also praised by Continental mathematicians: Johann Bernoulli was particularly complimentary about *De Sectione Rationis*, for example.[19] After this success, Halley prepared an edition of the *Spherics*, on the geometry of the sphere and its applications in astronomy, by the 1st-century Menelaus of Alexandria, translating it into Latin from Arabic and Hebrew manuscripts.[20] This work was printed at the time, but (for reasons now obscure) was not published until 1758, long after Halley's death.

During his early years as Savilian professor, Halley made many contributions to Oxford life. His reputation attracted distinguished foreign visitors to Oxford and, as befitted his professorship, he lectured on mathematics – for example, in the autumn of 1704, on the 'Geometrical Construction of Algebraical Equations; And the Numerical Resolution of the same by the Compendium of Logarithms'; these lectures were published in 1717 as an appendix to a new printing of the mathematical teacher John Kersey's *Elements of that Mathematical Art commonly called Algebra*, first published in 1673.[21] Indeed, Halley's interest in polynomial equations pre-dated his incumbency of the Savilian

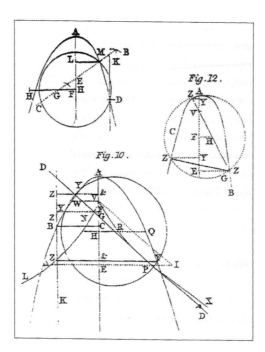

Diagrams from Halley's work on the roots of cubic and quartic equations, published in the *Philosophical Transactions* of the Royal Society in 1687.

chair by more than a decade, as shown by three papers delivered to the Royal Society (two in 1687 and one in 1694) on an iterative method for finding the roots of such equations.[22] These were well received by the Society, if the reaction of the naturalist Richard Waller is anything to go by:[23]

Mr Halley is daily entertaining us with new Discoveries and has now shown us a new short and easy method of extracting the Roots of all Equations.

Moreover, it was a diagram from one of his algebraic papers of 1687 that Halley held in the portrait shown earlier.

It was also around this time that Halley made his most profound and intellectually far-reaching contributions to astronomy, as well as further developing the Newtonian impact upon the University, already commenced by David Gregory. But his real contributions to astronomy during 1704–20 were made in his papers on the physical composition of deep space, variable stars, stellar distribution, and the proper motion of stars.[24] During his professorial years many of Halley's long-standing research interests, such as geomagnetism and planetary dynamics, came to maturity. It is truly remarkable how Halley continued to maintain a veritable outpouring of original ideas and discoveries in the physical sciences, in spite of the diverse demands placed upon him.

The Savilian professors' houses in New College Lane, with the observatory that was built on the roof on Halley's instructions and which his colleague David Gregory described as 'very convenient and indeed useful to the university, and what Sir Henry Savile did expect from them'.

Halley as Astronomer Royal

John Flamsteed died on the evening of 31 December 1719, after having served forty-four years as Britain's first Astronomer Royal. Neither he nor Halley had cared much for each other since their early interactions in Halley's youth and, as we have seen, Flamsteed had done undisputed mischief to Halley's career by sabotaging his application for the Savilian chair of astronomy in 1691. Halley, in turn, had seized some opportunities to get his own back, most especially in 1712 when he was prevailed upon by Newton to edit Flamsteed's observations as *Historia Coelestis in Libri Duo* (History of the Heavens in Two Volumes). Not only did Halley's editorial rigour leave much to be desired – the printed volumes contain many numerical errors – but he added a preface in which he attacked Flamsteed for sluggishness, secretiveness, and a lack of public spirit. This 'pirated edition' of Flamsteed's lifetime work drove home the final division between Halley and Flamsteed. It was therefore an enormous affront to the memory and achievement of Flamsteed when, early in 1720, Halley was appointed Astronomer Royal to succeed him. Flamsteed's widow, Margaret, was enraged and sold all of her husband's instruments (which were the Astronomer Royal's personal property) to ensure that Halley could never lay hands upon them.

Halley probably sought the vacant Astronomer Royal's job for the £100 annual salary that it carried, but also because he could hold it in tandem with his Oxford professorship.

Edmond Halley in later life.

By the pluralist standards of the Hanoverian age it was considered natural for a talented and energetic man to accumulate offices, and it was not necessary for Halley to resign his Oxford chair on account of his new appointment. Yet Halley, as in Oxford, did not treat Greenwich as a sinecure, but threw himself into a determined programme of research there. It says a great deal for his apparent health and vigour that he was offered this arduous job in the first place, bearing in mind that he was already 64 years old, and would be 69 before his new set of instruments was ready. More extraordinary was his declared intention of devoting his office to the observation of the motion of the lunar coordinates through a full eighteen-year cycle, necessary for the long-proposed method of finding longitude at sea from the Moon's position against the fixed stars. Most extraordinary of all, he succeeded in accomplishing this task and lived to be Astronomer Royal for nearly twenty-three years.

When Halley took up residence at Greenwich in 1720 he moved into an observatory with no instruments. Being shrewder and more courtly than Flamsteed, he persuaded the government to pay for them – a fortunate precedent for future holders of the office. Even so, he had to start from scratch, and with his generous £500 official grant he set about designing, ordering, and supervising the construction of what by 1725 would become a prototype set of meridian astronomy instruments. He was also fortunate in securing the services of George Graham FRS, the great clockmaker and foremost mechanician of the age, to construct them. Not only were the instruments produced of fundamental importance for Halley's own work, but their design set a precedent for all subsequent large observatory instruments from the 1720s to the early 20th century.[25]

Halley never obtained lunar coordinates that were sufficiently accurate to find longitude reliably or to win the £20,000 prize that had been established under the Longitude Act of 1714. This did not derive from any slackness or lack of zeal on his part; indeed, it was said that in eighteen years Halley never missed a single lunar transit that was visible at Greenwich – a formidable achievement for anyone, let alone someone entering his 80s. His failure derived from the fact that it was not then possible to measure the Moon's position against the fixed stars with sufficient accuracy, even on Graham's superlative instruments. Not until the 1760s, with further improvements in mechanics and precision instrument making, would it be possible to measure down to the one or two arc seconds that were necessary.

Nathaniel Bliss, Joseph Betts, and John Smith

Upon Edmond Halley's death in 1742, the clergyman Nathaniel Bliss succeeded to the Savilian professorship of geometry. Born in Gloucestershire in 1700 and educated at Pembroke College, Oxford, Bliss had since 1736 been Rector of St Ebbe's Church in Oxford, a stone's throw from his *alma mater*. As an undergraduate he had attended the lectures of the influential astronomer James Bradley, Halley's protégé, successor as Astronomer Royal, and Savilian Professor of Astronomy from 1721 until his death forty years later. Bradley's lectures did much in Oxford to establish the Newtonian conception of the universe, and it is probably these that inspired Bliss's interests in mathematics and astronomy – but like his Savilian predecessor, Bliss's main focus was on the latter. Indeed, he established his own observatory in Oxford by fixing instruments to the section of the old city wall adjacent to the Savilian professors' houses in New College Lane.[26]

Bliss is a tantalizing figure, in that he left few enduring achievements or records behind him. We know that he was elected to a Fellowship of the Royal Society in 1742, with Bradley as the first signatory on his election certificate,[27] and that he made observations of Jupiter's satellites in 1742 and of the bright comet of 1745. We also know that he had helped Bradley in some of his work at the Royal Observatory – for example, he

Nath.ᶜ Bliſs A.M. Profeſsor of Astronomy at Oxford. F.R.S Octʳ 1764 ætˢ 64.

(*Left*) Nathaniel Bliss (1700–64).
(*Right*) This drawing of Bliss was apparently scratched on an old pewter flagon during dinner by the astronomer George Parker FRS (later 2nd Earl of Macclesfield); it was turned into this image by the distinguished engraver James Caldwell. Notice that Bliss is incorrectly labelled as 'Savilian Professor of Astronomy'.

made observations of the 1761 transit of Venus when Bradley was too ill to do the work himself.[28]

Otherwise, there is little that we can say about him, although he was clearly a figure of considerable significance and energy in his day – not only was he a vigorous Oxford teacher, but he was also sufficiently weighty in national scientific terms to be elected Astronomer Royal in succession to Bradley in 1762. Sadly, however, and in contrast to Halley who worked for a further twenty-two years following his appointment as Astronomer Royal at the age of 64, Bliss died just two years into the job. As he had completed no major researches during that time, his Greenwich observations were not published until 1805, when Thomas Hornsby, Bradley's successor as Savilian Professor of Astronomy, included them as a supplement to his edition of Bradley's observations.[29] As a consequence, Bliss as Astronomer Royal has been somewhat overshadowed by both Halley and Bradley.[30]

We may suppose, based upon the advertisement for his lectures, that Bliss took his professorial duties seriously, lecturing to small groups on such topics as arithmetic, Euclid's *Elements*, plane and spherical geometry, conic sections, the use of logarithms and surveying instruments, and experimental philosophy. Not all of Bliss's auditors were enthusiastic about his teaching, however; the young Jeremy Bentham, who entered The

N. BLISS M.A. *Savilian Professor of Geometry*, proposes to explain the Elements of the most useful Mathematical Sciences, at his House in *New-College* Lane, in the following *Classes* or *Courses*.

I. ARITHMETICK *Vulgar* and *Decimal* with its Application to common Affairs, as well as to the other Parts of the Mathematicks.

II. The first *Six Books* with the *Eleventh* and *Twelfth* of EUCLID'S *Elements*.

III. ALGEBRA, wherein will be taught the Method of resolving the several kinds of *Equations*, illustrated by a great Variety of useful and curious Problems, as well *Arithmetical* as *Geometrical*.

IV. PLAIN TRIGONOMETRY, wherein will be shewn, the Construction of the *Natural Sines, Tangents* and *Secants*, and the Table of *Logarithms*, as well of the *Natural Numbers*, as of the *Sines, Tangents* &c. with the Use of the *Logarithmic Tables* in the Solution of the several Cases of *Plain Trigonometry*. To which will be added the *Practical* GEOMETRY, comprehending the Description and Use of *Instruments*, and the Manner of measuring *Heights, Distances, Surfaces*, and *Solids*.

V. SPHERICAL TRIGONOMETRY, with its Use in the Resolution of the most common Problems of the *Sphere*; together with the Method of Projecting the several Cases *Stereographically*. To which will be added the full Description and Use of both the Celestial and Terrestrial Globes and the Method of Solving the same Problems by them, which were solved by Spheric Trigonometry.

VI. The Elements of the CONIC SECTIONS, with the Demonstration of such of their Properties as are of most frequent Use; together with the Mensuration of the Superficial and Solid Content of the *Cone* and its *Frustums* and *Sections*. To which may occasionally be added the Method of Projecting the Sphere *Orthographically*, exemplified in the Construction of *Solar* and *Stellar Eclipses*.

It is proposed that the Number of *Scholars* in each of these *Classes* or *Courses*, be not less than *Six*, or more than *Ten*; to whom he will read three Days in a Week, and not less than an Hour each Day, explaining the Propositions, and illustrating them with Examples, and such Observations as the Matter shall require until the Company apprehend and understand it: And each *Person* shall have full Liberty to propose such *Doubts* or *Scruples* as he pleases.

For the *Text* to be explain'd, and to give Occasion for necessary *Digressions*, a printed Book will be used, if there be any that is proper: in other Cases every *Scholar* shall have Liberty to transcribe his *MS. Notes*, if he pleases.

It is computed that any one of these *Classes* or *Courses* will require about *three Months*; and any *Gentleman* may go through any *one* or *all* of them as he pleases, paying *two Guineas* at the Beginning of each Course, and *half a Guinea* more for every *Month* the Course shall continue longer than *three*.

Advertisements for the mathematical lectures of Nathaniel Bliss, showing a spread of topics that remained staples of mathematics teaching in Oxford throughout the 18th century.

Queen's College in 1760, contrasted Bliss's practical and theoretical abilities in a letter to his father of March 1763:[31]

We have gone through the Science of Mechanics with Mr. Bliss, having finish'd on Saturday; and yesterday we begun upon Optics; there are two more remaining, viz: Hydrostatics, and Pneumatics. Mr. Bliss seems to be a very good sort of a Man, but I doubt is not very well qualified for his Office, in the practical Way I mean, for he is oblig'd to make excuses for almost every Experiment they not succeeded according to expectation: in the Speculative part, I believe he is by no means deficient.

We return below to the Oxford mathematics teaching of the 18th century.

We see from Bentham's letter to his father that Bliss continued to teach in Oxford, even after his appointment as Astronomer Royal. Moreover, following his death, Bliss's enterprising widow arranged what one assumes to have been a continuation of his popular lectures, delivered by Thomas Hornsby. Elizabeth Bliss was clearly the driving force behind a special lecture entitled 'Electrical Experiments for the Entertainment of Ladies and others', given in the Bodleian Library tower above the Schools quadrangle on 21 May 1765, for her name heads the published prospectus as organizer, taking precedence over Professor Hornsby who merely delivered it, and stipulating that admission would be by ticket for half a crown. One assumes that the ladies who attended the lecture were well off, for in 1765 half a crown would have maintained a labourer's entire family for a couple of days.[32]

If Bliss has left little trace of himself in the historical record, then his two Savilian successors, Joseph Betts and John Smith, are even more tantalizing – the former in particular. As with both Halley and Bliss, mathematics was not the primary focus of either successor: Betts was another astronomer, while Smith was a physician.

Even Joseph Betts's entry in the published register of Oxford alumni has little to say about him: he hailed from Deptford in Kent, and entered University College in 1736 at the age of 18; he earned his BA degree in 1740, his MA degree in 1743, and was elected a Fellow of his College in that same year.[33] Indeed, it is in this last capacity, rather than in his later occupancy of the Savilian chair, that we find fleeting references to him – for example, as a subscriber to William Gardiner's *Tables of Logarithms* of 1742,[34] and as author of a 1744 letter to the *Philosophical Transactions* concerning observations of a comet, perhaps made in collaboration with Bliss.[35]

Despite having prominent backers, Betts applied unsuccessfully for the post of Savilian Professor of Astronomy upon Bradley's death in 1762 (losing out to Thomas Hornsby), and for that of Astronomer Royal following Bliss's death in 1764 (losing out to Nevil Maskelyne).[36] His application for the Savilian chair of geometry was more successful, but he occupied the post for only a year and left little trace of his time as professor. He was buried in the chapel of University College in January 1766.

Betts's successor, John Smith, stands out in this century of astronomers, and indeed in the history of the Savilian chair of geometry more generally, as seemingly having had no mathematical credentials whatsoever. Born in Ayrshire, he studied first at the University of Glasgow, and then at Balliol College, Oxford, receiving his BA degree in 1748, and working eventually towards to a medical doctorate in 1757; around this time, the University's anatomy lecturer, Nathan Alcock, departed for private practice in Bath, and Smith took over that role. He subsequently lectured in chemistry as well, at least until his appointment as Savilian professor in 1766.[37]

How active Smith was in this latter post is open to question. The cleric Alexander Carlyle noted in his autobiography that Smith 'taught mathematics with success in Oxford', but it is not clear whether this assertion was derived from direct knowledge or was an assumption based on Smith's title.[38] Smith became physician at the Radcliffe Infirmary in Oxford upon its opening in 1770,[39] and by 1784 he was practicing medicine in Cheltenham; at this point his eventual Savilian successor, Abraham Robertson, was certainly deputizing for him in his mathematical lecturing duties.[40] We also have evidence of an earlier deputy, Smith's fellow medic William Austin, who published a short *Examination*

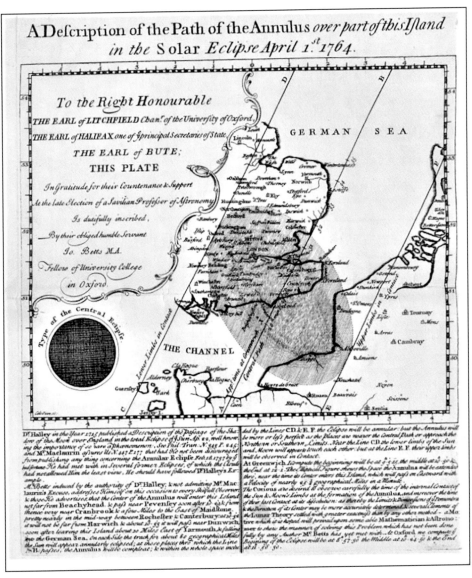

Joseph Betts's engraving of the path of the annular solar eclipse of 1 April 1764, with a dedication to the supporters of his unsuccessful application for the Savilian chair of astronomy.

of the First Six Books of Euclid's Elements as 'a small effort towards a plan for facilitating the study of geometry', dedicating it to Smith.[41] Meanwhile, from Cheltenham, Smith wrote a short text on the 'use and abuse' of the waters there, proudly displaying the title of Savilian Professor of Geometry on its opening page,[42] but his only lasting contribution

John Smith's book on the waters at Cheltenham, displaying his title of Savilian Professor.

to the chair seems to have been the stable and the tenement that he erected behind the Savilian professors' houses.[43]

Mathematics among the astronomers

In surveying the activities of the Savilian Professors of Geometry throughout the 18th century, we have painted a picture of a succession of scholars whose interests, in most cases, spanned both mathematics and astronomy. Indeed, in line with earlier comments about the interchangeability of the Savilian chairs, the same may be said of the astronomy professors, who provided another source of mathematics teaching in Oxford throughout this period.

Halley's collaboration with David Gregory, Savilian Professor of Astronomy from 1691 to 1708, in the preparation of his edition of Apollonius's *Conics* has already been mentioned. David Gregory was a nephew of the great Scottish mathematician James Gregory, and went on to develop many of his uncle's ideas. He was also an enthusiastic

The memorial to David Gregory (1661–1708) in the University Church.

teacher in his own right, whom Newton strongly and successfully recommended for Oxford's astronomy chair when it fell vacant in 1691; this, we recall, was the occasion of Halley's unsuccessful application, and one wonders whether a Newtonian recommendation for a rival candidate had as much to do with this failure as Flamsteed's sabotage.

On his arrival in Oxford, Gregory became the University's first active resident promoter of Newtonian ideas. His inaugural lecture on 21 April 1692 sketched the history and methods of celestial physics, with particular emphasis on Newton's work; indeed, the second half of the lecture, a catalogue of results established in the *Principia*, must have been the fullest public account of Newton's work yet heard in Oxford.

Certainly, by the early years of the 18th century, Gregory was actively (but tactfully) advertising for mathematics students, under the heading of 'Dr. Gregory's Method for Teaching Mathematicks':[44]

Without discouraging any other person in the University that teaches or intends to teach Mathematicks; at the desire of persons of note, he undertakes to teach the different parts

Oratio Inauguralis
a Davide, Gregorio M: D: Astronomiæ pro-
fessore Saviliano, in Auditorio Astronomico
Oxoniæ, Habita vigesimo primo die Mensis Aprilis
Anni 1692, quum publicam [crossed out] professionem
auspicatus est.

Insignissimo Domino Vice Cancellario.
Reliquiisq. Academiæ Procores venerandi.
Academici Doctissimi.

Professionis Astronomicæ munus in Academia
Oxoniensi suscepturum decet in ipso principio, grati
Oratione Excellentissimi viri D. Henrici Savili
Memoria litare, qui Astronomiis insomnibus illis
portavum mundi vigilibus hæc otia fecit, et il
lustrissimis professorum Savilianorum electoribus
devinctissimum, me profiteri, qui ea nobis quoque
propria esse voluerunt.

David Gregory's inaugural lecture as Savilian
Professor of Astronomy asserted the value of
geometry in understanding the world, and
promoted the improvement of natural
philosophy along these lines.

and sciences of the Mathematicks (by way of Colleges and Courses) after the manner following.

If any number of Scholars desire him to explain to them the Elements (or any of the Mathematical Sciences if they are already acquainted with the Elements) He will allow that Company such a time, as they among themselves shall agree upon, not less than an hour a day, for three days in a week; in which time he will go through the said Science; explaining the propositions, and illustrating them with examples, operations, experiments, or observations, as the matter shall require, until the Company apprehend and understand it: And there shall be full liberty to every Person of the Company to propose such doubts and scruples as he pleaseth.

And because some may be desirous to give an account of their proficiency; for their own satisfaction and that of their friends, he will once a week examine such as shall signify that they are willing to be examined.

Gregory went on to suggest seven mathematical courses, covering parts of Euclid's *Elements*, plane geometry with practical applications, algebra (including the solution of Diophantine problems), mechanics, optics, the principles of astronomy (including spherical trigonometry), and the theory of the planets – in short, a list of topics that was not very different from that offered by Bliss half a century later. It is not clear, however, how many students took up Gregory's offer, and his early death in 1708 necessarily put an end to any such plans.

Gregory was succeeded as Savilian Professor of Astronomy by the geodesist and instrument-designer John Caswell, whose visible mathematical credentials included *A Brief (but Full) Account of the Doctrine of Trigonometry* of 1685, which was appended to John Wallis's *Treatise of Algebra*, and his contribution to the solution of a problem of quadrature.[45] Caswell's occupancy of the chair was short-lived, however: he died in 1712, and was succeeded by another strongly mathematically inclined figure, John Keill.

Keill was a former Edinburgh pupil of Gregory and had followed him to Oxford, where he subsequently served in various lecturing capacities before succeeding to the Savilian chair.[46] He had previously applied for it upon Gregory's death in 1708, as well as for Oxford's Sedleian chair of natural philosophy in 1704, each time without success, and his supporters asserted that there had been a Whig Low Church conspiracy against him: unusually for a 17th-century Scot, Keill was a High Churchman. In 1701, Jonathan Swift wrote to a friend:[47]

You know, I believe, that poor Dr Gregory is dead, and Keil sollicites to be his successor. But Party reaches even to Lines and Circles, and he will hardly carry it being reputed a Tory, wch yet he wholly denyes.

Keill argued against the 'atheistic' philosophies that were emerging from the Continent, and emphasized the compatibility of Newtonianism and Christianity. He was especially interested in Newtonian explanations for the coherence of matter and its relation to experimental philosophy. It was Keill who introduced the teaching of Newtonian ideas to Oxford through courses on experimental philosophy in the 1690s. According to the later testimony of the natural philosopher and fellow Newtonian, John Theophilus Desaguliers,[48] Keill

was the first who publickly taught Natural Philosophy by Experiments in a mathematical Manner: for he laid down very simple Propositions, which he prov'd by Experiments, and from those he deduc'd others more compound, which he still confirm'd by Experiments; till he had instructed his Auditors in the Laws of Motion, the Principles of Hydrostaticks and Opticks, and some of the chief Propositions of Sir Isaac Newton concerning Light and Colours. He began these Courses in Oxford . . . and that Way introduc'd the Love of the Newtonian Philosophy.

During his decade as Savilian Professor of Astronomy, Keill lectured and published both on astronomy and on mathematics, the latter in the form of a treatise on plane and spherical trigonometry.[49] He was a member of a group of early 18th-century figures (that also included Gregory and Desaguliers) who brought important pedagogical gifts to Oxford which enabled them to promote the Newtonian philosophy in a form that could be comprehended by mainly non-scientific audiences. This must have done a great deal to assure its acceptance among the educated laity, which constituted the

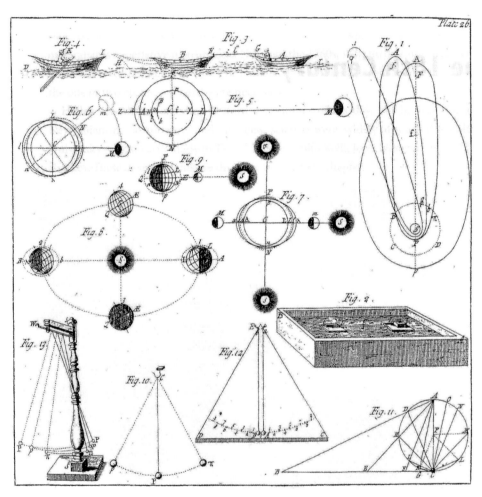

John Theophilus Desaguliers (1683–1744) won fame as a lecturer on experimental philosophy, first in Oxford and then in London. His great talent lay in the high-level popularization of Newtonian physics, and here, in his *Course of Experimental Philosophy* of 1734, he gave several plausible demonstrations of Newton's third law of motion, involving boats, magnets, tides, and pendulums.

academic and undergraduate body. When one remembers the overall Christian context of their teaching, they must have played an important part in showing that the new science was compatible with Anglicanism, by demonstrating the logic and order of the Creation.

The incorporation of the Newtonian philosophy into the cultural fabric of the Georgian intelligentsia owed a great deal to Gregory, Keill, and men like them. The lectures of James Bradley, Keill's successor as Savilian professor and Halley's as Astronomer Royal, which we have already mentioned in connection with Nathaniel Bliss, continued in this tradition. Bradley was an inspired and gifted teacher of the mathematical sciences who developed a significant following: he eventually averaged fifty-seven students per course, who paid him three guineas (£3.15) per head, and among his surviving papers in the Bodleian Library are the names of over one thousand individuals. Also surviving in Oxford are the detailed notes of a student who attended Bradley's 1747 lectures on experimental philosophy. This student notebook gives a remarkable insight into the teaching of physics in 18th-century Oxford, and shows how a scientist of international standing presented the subject to a non-specialist audience.[50]

While Bradley was finding success in communicating a largely non-mathematical version of Newtonian ideas, the situation with regard to mathematics teaching in Oxford

James Bradley (1692–1762), Keill's successor as Savilian Professor of Astronomy and Halley's successor as Astronomer Royal.

A syllabus for either Bradley's course of lectures in the 1740s or the similar one by his 1720s predecessor, John Whiteside.

was considerably less auspicious. Following the death of John Keill in 1721, the Provost of Oriel College, George Carter, commented in a letter to William Wake, Archbishop of Canterbury:[51]

Mathematics are studyed, Your Grace knows very well, but little amongst us, especially by such as I should be glad to recommend to this Place.

At around this same time concerns were being raised about the suitability of some of Oxford's incumbent professors – though often by figures with an axe to grind, such as Nicolas Amhurst, who had been ejected from his Fellowship at St John's College in 1719:[52]

I have known a profligate *Debauchee* chosen Professor of *Moral* Philosophy; and a Fellow, who never looked upon the Stars, sober in his life, Professor of *Astronomy*.

This last-cited professor was apparently John Keill, though it is not known whether the accusation had any basis. Indeed, an analogous charge could have been levelled at this stage at Halley as Savilian Professor of Geometry: having become Astronomer Royal in 1720, he had moved to Greenwich and had largely turned his back on Oxford teaching and on any mathematics that did not bear directly on astronomy.

Mathematics teaching in Oxford

The reasons for the relative neglect of mathematics in 18th-century Oxford are complex. They relate in part to the place of (Newtonian) mathematics within the wider political landscape of the age. Oxford was identified throughout this period with High Church and Tory opposition to the Hanoverian dynasty. Newtonianism, as a general approach to understanding the world, became increasingly identified with the Whig oligarchy, and it may be that political and theological tensions contributed to the way that Oxford mathematical activity died away over the course of the century, particularly in the absence of the former strong personalities such as Wallis, Gregory, and Keill.

Historical scholarship has tended to present 18th-century Oxford as having descended into a decadent slumber more generally, although this point of view has been somewhat overstated. Mathematics may have languished at this time, but other fields of scholarship – particularly those in the sciences – continued to flourish. This was the period of the foundation of various institutions funded by, and named for, the Radcliffe estate: the Camera (to house a scientific library), the Infirmary, and the Observatory. The biomedical sciences also benefited from the establishment of the Lee Anatomy Laboratory at Christ Church, which served the whole University, and from the foundation of a number of chairs and readerships. Even mathematics teaching did not vanish from the University altogether: we see evidence of its continuation, for example, in the reissuing later in the century of Keill's student treatise on trigonometry, and of his edition of Euclid's *Elements*.[53]

Further evidence of the availability of mathematics teaching in 18th-century Oxford as part of a broader liberal education, perhaps only at a low level, may be found scattered throughout the memoirs and letters of students of the period; for instance, we have already recorded the views of Jeremy Bentham on the lectures of Nathaniel Bliss. Around the same time, Charles James Fox, later a prominent Whig politician, but then an undergraduate at Hertford College, wrote to a friend:[54]

I read here much, and like vastly (what I know you think useless) mathematics. I believe they are useful, and I am sure they are entertaining, which is alone enough to recommend them to me.

However, Fox's remark must be set alongside one made by his nephew, Lord Holland, many years later: after asserting his uncle's aptitude 'for all branches of mathematics', Holland noted further that:[55]

I have often heard him regret that he had applied so little to them; and ascribe his neglect of them to the superficial manner in which they were taught at Oxford.

This superficiality may have been due to the laid-back attitude of Fox's tutor, William Newcome, who does not appear to have shown much concern over the amount of time that Fox spent away from Oxford, and who may even have delayed the geometrical studies of fellow students until Fox's return.[56] The pace of mathematics teaching a short distance away in Balliol College was similarly sedate: in the early 1770s a pamphlet was circulated that criticized a mathematics lecturer there, one Samuel Love, for having taken a whole term to reach only Proposition 7 of Book I of Euclid's *Elements*.[57]

Elsewhere in Oxford, mathematics formed a part of the disciplinary system: as punishment for a now-forgotten offence, one student at Christ Church was ordered to master several books of Euclid's *Elements*, as well as to work through all the examples in the first part of Colin Maclaurin's *Treatise of Algebra*. Punitive mathematics aside, surviving reading lists at Christ Church indicate that little mathematics was read by undergraduates there up to 1760, but thereafter several versions of Euclid's *Elements*, as well as the various editions of Maclaurin's *Algebra*, begin to appear.[58] Nevertheless, the image of Oxford generally, as the wrong English university for those interested in mathematics, persisted well into the 19th century.[59]

Robertson and Rigaud

To return to the Savilian professors of geometry, we note that the death of John Smith in 1797 marked a watershed, in that it enabled the chair to be filled once again by an active mathematician: Smith's sometime deputy, Abraham (or Abram) Robertson. His eventual successor, Stephen Peter Rigaud, was also no stranger to mathematics. However, astronomy remained a major interest for both Robertson and Rigaud – indeed, both ultimately exchanged the geometry chair for that of astronomy.

Robertson was born in Berwickshire in 1751 and worked his way up from humble origins. Around the age of 24, he moved to London in the hope of securing a position with the East India Company, but when this plan failed he relocated to Oxford, where he attempted (unsuccessfully) to make a living by opening an evening school for teaching mechanics, and subsequently worked as an assistant to an apothecary.

It may have been in this latter setting that he first encountered John Smith, whose patronage enabled Robertson to enter Christ Church as a student in 1775, eventually becoming a chaplain there in 1782. Within two years, as we have seen, he was deputizing for Smith as a mathematical lecturer, a role that gave rise to his earliest publications, such as his short clarification of Definition 5 of Book V of Euclid's *Elements*, 'printed for the use of those who attend the public mathematical class'.[60]

A celebrated Latin treatise on conic sections followed in 1792, and in that same year Robertson saw to publication Giuseppe Torelli's edition of the works of Archimedes.[61] Both of these volumes were cited in connection with Robertson's 1795 election to a Fellowship of the Royal Society, 'on account of his literary attainments and diligence in the pursuit of Science'.[62] An English treatment of conic sections followed in 1802, and a further abridged version sometime later.[63] During these years, Robertson also contributed papers on a new proof of the general binomial theorem to the *Philosophical Transactions*,[64] although he soon found himself embroiled in a dispute as to whether his proof had already been given by Leonhard Euler in 1774.[65]

Robertson appears to have been a successful and popular lecturer who was attentive to the needs of his students: his publications on Euclid and conic sections were written expressly for their benefit. An obituary of Robertson[66] observed that his

manner of lecturing was deliberate and perspicuous; and he was always ready to assist and encourage the students who attended him; he frequently lent them his papers to examine at their leisure . . .

One such student, George Chinnery of Christ Church, wrote to his mother almost every day during his years in Oxford, and these letters have survived to give us an impression of mathematics teaching during the first decade of the 19th century. With regard to Robertson in particular, Chinnery reported the following in February 1810:[67]

I rejoice most amazingly in being able to tell you that the time for the opening of Robertson's Newtonic lecture is fixed; it will be on the 15th. I had a long conversation with him yesterday morning at his house on the subject, and he read the prospectus of the course of lectures

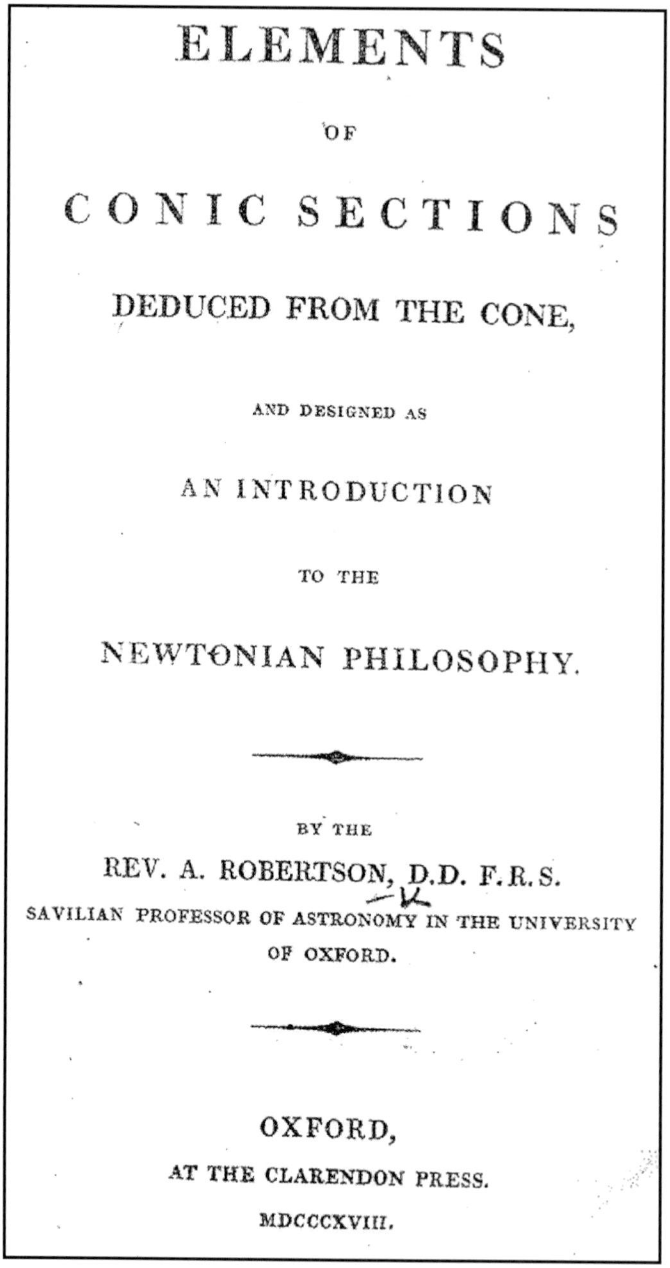

ELEMENTS

OF

CONIC SECTIONS

DEDUCED FROM THE CONE,

AND DESIGNED AS

AN INTRODUCTION

TO THE

NEWTONIAN PHILOSOPHY.

———◆———

BY THE

REV. A. ROBERTSON, D.D. F.R.S.

SAVILIAN PROFESSOR OF ASTRONOMY IN THE UNIVERSITY
OF OXFORD.

———◆———

OXFORD,

AT THE CLARENDON PRESS.

MDCCCXVIII.

The third version of Robertson's treatise on conic sections, produced after his move to the Savilian chair of astronomy.

to me; among other things, he advised me before the 15th to study the beginning of the Principia (which is that part of Newton that forms the object of our attention) by myself; and this I mean to do.

And two weeks later:

Robertson's 4th lecture took place today; there is one difficulty attending them, from their being public, that if in a demonstration (which he does upon a slate before us) there should be any point not immediately clear, one cannot stop him & beg that he will begin over again; this however signifies but little; it obliges to unremitting attention during the hour & a half which is the time allotted to the lecture, and complete really undivided attention will prevent any part of his demonstration being unintelligible; he is a delightful beautifully slow, beautifully clear lecturer, 1000000000 times better than Lloyd. I like him much.

The 'Lloyd' mentioned here was Charles Lloyd, subsequently Regius Professor of Divinity and Bishop of Oxford.

Already during his years as geometry professor, Robertson had demonstrated an interest in astronomy by submitting a paper on the precession of the equinoxes to the *Philosophical Transactions*,[68] and was also responsible for the publication of the second volume of Bradley's *Observations*. When Thomas Hornsby died in 1810, Robertson succeeded him as the Radcliffe Observer,[69] and it is perhaps for this reason that Robertson was then permitted to take the sideways step from one Savilian chair to the other. During the previous two centuries, the geometry and astronomy professorships had been largely interchangeable, but Robertson's translation from one to the other is perhaps an indication that, by the beginning of the 19th century, the chairs were beginning to be tied more firmly to their titular subjects. Robertson's sparse publications certainly lean more towards the astronomical than the mathematical in the years following his move to the astronomy chair.[70]

Robertson's successor (in both chairs) was Stephen Peter Rigaud, a descendent of French Huguenots and a member of a family with firm astronomical credentials: his father, grandfather, and uncle had all carried out work at the King's Observatory at Kew, and Rigaud himself eventually succeeded to the position of Observer there in 1814. In 1791 Rigaud entered Exeter College, Oxford, at the age of 16, and followed the traditional path of the BA and MA degrees and a college Fellowship – the last of these in 1794, even before he had completed his degrees. He engaged in tutorship within Exeter College, and from around 1805 he deputized for the ailing Thomas Hornsby as lecturer in experimental philosophy, a post that he then took on formally upon Hornsby's death in 1810. In that same year, Rigaud took over from Robertson as Savilian Professor of Geometry, and would similarly take the sideways step into the astronomy chair after Robertson's death in 1827, at which point Rigaud also became the Radcliffe Observer.

Over the course of his career, Rigaud filled various other posts within the University, including Public Examiner (following the institution of a new examinations system at the beginning of the 19th century) and Delegate to the University Press. An 1839 obituary in *The Oxford Herald* noted that, since his arrival at Exeter College in 1791, he 'had never been absent from Oxford so much as a single year during the period which has since elapsed, a little short of half a century'.[71] In 1828 Rigaud, together with the geologist

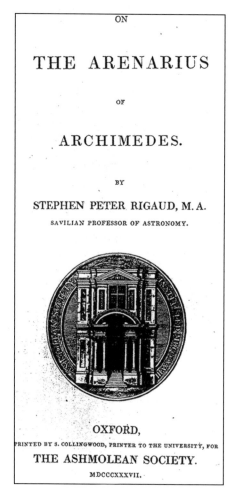

ON

THE ARENARIUS

OF

ARCHIMEDES.

BY

STEPHEN PETER RIGAUD, M. A.

SAVILIAN PROFESSOR OF ASTRONOMY.

OXFORD,

PRINTED BY S. COLLINGWOOD, PRINTER TO THE UNIVERSITY, FOR

THE ASHMOLEAN SOCIETY.

MDCCCXXXVII.

(*Left*) Stephen Rigaud's *On the Arenarius of Archimedes*, written while he was Savilian Professor of Astronomy.
(*Right*) A silhouette of Stephen Peter Rigaud (1774–1839).

William Buckland, founded the Ashmolean Society, an Oxford scientific club that met to read and discuss papers on a variety of subjects. Rigaud's own contributions to the society were very varied and included 'Remarks on the proportionate quantities of rain at different seasons in Oxford', 'Account of some early proposals for steam navigation', and 'On the *Arenarius* of Archimedes'. The last of these signals Rigaud's interest in the history of science, a topic to which we return shortly.

In June 1805 Rigaud was elected to a Fellowship of the Royal Society and would eventually become its Vice President. On his election certificate, signed by Robertson among others, he was cited as 'a gentleman well versed in Philosophical Learning and especially in the several branches of the Mathematics'.[72] The *Oxford Herald* obituary noted above spoke of his 'correctness of judgment, which qualified him for mathematical pursuits'.[73] However, Rigaud does not appear to have published anything of a strictly mathematical nature, nor has he left much trace of his mathematical teaching. In contrast, a *Syllabus* of his lectures in experimental philosophy has survived,[74] and we know that during the 1820s the average attendance of this course was around forty students, bringing Rigaud substantial fees.[75]

The meridian circle installed by Rigaud in the Radcliffe Observatory.

Like so many of his immediate predecessors in the geometry chair, Rigaud's main interest was in astronomy. He began his observational career as an undergraduate and continued to make measurements, both meteorological and astronomical, throughout his life. In the 1830s he oversaw the installation of the Radcliffe Observatory's first major new instrument, a six-foot mural circle by Thomas Jones of London, which increased the accuracy of the Observatory's meridian astronomy.[76] However, Rigaud was to be the last figure to hold both the position of Savilian Professor of Astronomy and that of Radcliffe Observer. Upon his death in 1839, Rigaud was succeeded as Observer by Manuel Johnson, who was not appointed to the Savilian chair, possibly for religious reasons. Instead, it went to the clergyman George Johnson, a sometime mathematical examiner within the University, but a man with no observational experience. The blurring of the lines between the Savilian professorships of geometry and astronomy persisted even at this late stage.

Looking back

Early in this chapter, we remarked upon Edmond Halley's interest in the history of science, and on the way in which this fitted with the ethos of the chairs established by Henry Savile. As we reach the end of this century of astronomers, it is satisfying to observe, therefore, that both Robertson and Rigaud also shared such an interest – Rigaud, in particular. Robertson's connection with Torelli's edition of the works of Archimedes has been noted, and a majority of the works published by Rigaud in the final two decades of his life concerned the history of science in England.

During the 1780s the mathematical manuscripts of Thomas Harriot (c.1560–1621) were rediscovered at Petworth House in Sussex, the southern home of the Percy family: one member of this family, Henry Percy, 9th Earl of Northumberland, had been Harriot's patron for many years. The papers were passed to Oxford University Press with the recommendation that they be prepared for publication, whereupon the Press sought the opinion of Robertson, who was then acting as Smith's Savilian deputy.

Unfortunately, Abraham Robertson took the view that the papers were in no state to be published, and that in any case they were already of marginal scientific interest. However, upon a re-examination of the papers some years later, Rigaud disagreed.[77] Later described as 'the most historically sensitive mathematician to have studied them since Harriot's time',[78] Rigaud set himself the task of editing the papers for publication, and presented some of his results on Harriot's astronomical observations – specifically, his discovery, several months before Galileo, of the moons of Jupiter – to the Royal Society in 1832, and also at the meeting of the British Association for the Advancement of Science in Oxford in the same year.[79] Unfortunately, Rigaud's edition never appeared: the work was cut short by his death in 1839.[80]

Similarly unfinished was Rigaud's projected new edition of Pappus's 4th-century *Mathematical Collections*, although his manuscript notes survive in the Bodleian Library.[81] Otherwise, Rigaud's main historical interests were in the lives and work of the mathematical and astronomical men of the preceding 150 years, including Newton and Bradley.[82] Indeed, Rigaud's work on the former – on the *Principia*, in particular – sparked much of the subsequent Newtonian historical scholarship of the 19th century. Perhaps Rigaud's most famous work in this direction, published posthumously, was his *Correspondence of Scientific Men of the 17th Century*, featuring almost a thousand pages of transcribed letters, which Rigaud hoped thereby to preserve for posterity.[83]

Moreover, Rigaud's natural concern for astronomical instruments intersected with his interest in the history of science in the form of an article on the instruments employed by Halley as Astronomer Royal.[84] Indeed, Rigaud's historical interest in his Savilian predecessor enables us to conclude this chapter by bringing it full circle. Rigaud published some brief particulars of Halley's life, as well as an account of his cometary observations,[85] and planned to produce a longer biography of Halley, as 'a duty, to rescue his memory'.[86] Unfortunately, as with some of Rigaud's other historical writings, this was never completed.

Henry Smith, Savilian Professor of Geometry from 1860 to 1883.

CHAPTER 4

Baden Powell and Henry Smith

KEITH HANNABUSS

The years 1827–83, spanning the professorships of Baden Powell and Henry Smith, marked a period of great change in Oxford University. In the 1820s the University had been mainly a seminary for the Anglican clergy and a finishing school for the gentry. Reforms were introduced gradually, many as a result of Royal Commissions in 1850 and 1877, and by the 1880s Oxford had recognizably become a modern university. Although Powell was more of a scientist than a mathematician, he introduced important innovations that improved the status of mathematics within the University. His successor, Henry Smith, was both a reformer and a mathematician with an international reputation.

For the first 120 years the Savilian professorships of geometry were held by men of distinction, but in the later 18th century the short tenure of Joseph Betts and the election of the absentee physician John Smith broke the spell. Their successors, Abraham Robertson and Stephen Rigaud, were more able and conscientious than John Smith, but they were relatively minor figures in British mathematics. Nor was this malaise confined to mathematics, for the University had descended into a kind of torpor, from which it eventually recovered during the 19th century as it gradually transformed itself from an Anglican seminary and finishing school for the gentry into a recognizably modern university.

Keith Hannabuss, *Baden Powell and Henry Smith*. In: *Oxford's Savilian Professors of Geometry*. Edited by Robin Wilson, Oxford University Press. © Oxford University Press (2022). DOI: 10.1093/oso/9780198869030.003.0004

The recovery within mathematics owed much to Baden Powell and Henry Smith, successive Savilian professors of geometry for 55 years from 1827 to 1883. Baden Powell introduced reforms which created clearer career paths for mathematicians in the University, as well as being a respected member of the country's mathematical community, whilst Henry Smith was an internationally recognized mathematician of distinction. Both campaigned for University reform, and many changes came from the recommendations of the two Royal Commissions of 1850 and 1877, whose members included Powell and Smith, respectively.

Changes in the city and the university

The University reforms should also be seen in a broader context. In the 1830s the city of Oxford was still largely contained within the area encompassed by, or close to, its medieval walls. After the middle of the century the new developments of Park Town, Norham, and Walton Manors gradually pushed that boundary of the city ever further north,[1] but in the 1850s the map still showed few changes other than the new University Museum in the north, and the two new railway lines to the west.

When the Great Western Railway reached Oxford in 1844 it provided fast access to London and to Bristol, and it was soon joined by the London and North Western Railway in 1851, providing fast routes to Birmingham and Cambridge. Previously, the quickest way to reach the capital had been the express coach which took six hours to complete the journey, but that shrank to two hours and twenty minutes by rail. One unexpected consequence of the new connection was the discovery that one express coachman had a wife and family at both ends of his journey.[2] More significantly, it now became possible for academics to attend meetings of the Royal Society and other learned bodies and return to Oxford on the same day.

Until around 1860 the number of students matriculating at Oxford University each year stayed close to 400, but then it increased steadily to around twice that figure. At matriculation all students had to sign the Thirty-Nine Articles of the Church of England;[3] this meant that Catholics, Nonconformists, other Dissenters, and Jews could not study at Oxford. Cambridge University was slightly more relaxed, allowing students to postpone signing the Articles until graduation, and allowing otherwise excluded groups to study there but not to take their degrees. In 1855, groups from Oxford (including Baden Powell) and Cambridge (including James Joseph Sylvester) petitioned parliament to repeal these religious tests, but it took another sixteen years until the tests were finally abolished in 1871.[4]

In 1800 Oxford followed Cambridge in introducing Honour Schools which had a more rigorous written examination, rather than the cursory oral examination, before graduating with a BA degree in the traditional Pass Schools. Undergraduates could study for Honours in Classics (Literae Humaniores) or Mathematics (Disciplinae Mathematicae et Physicae); the physics in the Latin name was mainly applied mathematics and theoretical physics. Most colleges had both Pass degree and Honours degree students, but Balliol, Christ Church, and Oriel insisted that all students study for Honours. For a few years in the first decade, Classics was compulsory and Mathematics optional, but thereafter students could take either or both subjects, provided that they had demonstrated basic skills in the less demanding Responsions examinations.

The vast majority of undergraduates took Honours in Classics. Mathematics attracted only a minority of students, and most of these took Classics as well. In 1850 the University introduced new Honour Schools of Jurisprudence with Modern History, and Natural Science. In the early 1870s a new School of Sacred Theology was created, and Law and Modern History were separated into two Schools.

Although Classics initially lost numbers when the new arts courses were introduced, it remained the most popular choice, being usually at least twice as popular as any other subject. In the 1840s Mathematics was slowly becoming more popular, but after the 1850 reforms its numbers stayed relatively constant, whilst numbers in Natural Science gradually increased to a comparable level. In contrast, there were steady increases in those graduating in the new Arts Schools of Jurisprudence, Modern History, Sacred Theology, and the joint School of Jurisprudence with Modern History. In 1825 only 20 per cent of those matriculating had graduated with an Honours BA degree, but by 1885 the proportion was nearer to 45 per cent.

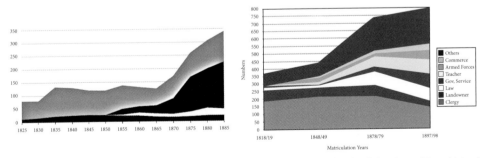

(*Left*) The numbers of students in Oxford's Final Honours Schools; from the top: Classics, Other Arts subjects, Natural Sciences, Mathematics.
(*Right*) Careers of Oxford graduates in the 19th century.

The most popular career after graduation was the clergy: by 1850 nearly half of all Oxford graduates entered the Church, and the figure was slightly higher at Cambridge. Next came the graduates from wealthy landowning families, who went back to manage their estates. In the course of the century these two groups came to represent smaller proportions of the graduates, while the professions, particularly in law, teaching, the civil service, and the armed forces, became more popular.

Most college Fellows had to be ordained within seven years of taking up their Fellowship, or to relinquish it at the end of that period, and were required to be celibate. Most colleges had 'livings', churches to which ordained Fellows wishing to marry could be 'presented' as clergymen, and those livings were mostly in remote rural parishes where research and scholarship could be continued alongside pastoral duties.

This created an obvious career progression for able students: obtain a good Honours degree, get yourself elected to a Fellowship (which were mostly decided by examinations), study for ordination and be ordained as a deacon and then a priest, and then ask your college to present you to one of its parishes, so that you could marry:

Student → Fellow → Ordination → Clergyman in a college living.

One general advantage of this was that college Fellowships were always available for new graduates as older academics left for marriage and the Church; in the mid-1830s the oldest Fellow at Balliol was only 26.

Baden Powell

This was almost exactly the career path chosen by Baden Powell. Baden was his Christian name, and the name was hyphenated as Baden-Powell only after his death; Robert Baden-Powell, the founder of the Boy Scouts, was his sixth son by his third wife.

Baden Powell matriculated at Oriel College in 1814 and, unusually, took Mathematics Honours only, gaining a first-class degree. He was then elected to a Fellowship at Oriel, was ordained, and then moved to be curate of Midhurst in West Sussex, before becoming vicar of Plumstead in south-east London. (This last arrangement was unusual, in that the right of presentation in Plumstead belonged to Powell's family, rather than to his college.) While in these parishes he pursued his research into the polarization of light and the infra-red spectrum of the Sun.

In 1824 Baden Powell was elected to a Fellowship of the Royal Society, and three years later he was elected as Savilian Professor of Geometry. (Charles Babbage, the computer pioneer, had been approached for the Savilian chair, but after much dithering did not

Two pictures of Baden Powell (1796–1860).

apply.) Powell's election seems to have owed less to his scientific credentials than to his membership of the Noëtics, a group on the liberal wing of the Church.

On his return to Oxford, Powell received rather a shock on discovering the small numbers and low standards of those studying mathematics; indeed, in some years no students took his courses. Another problem arose from the Honour Schools themselves, in that good students no longer found time to attend science lectures out of general interest, as they had done previously.

Baden Powell was advised not to give an inaugural lecture, because it would not attract an audience, so instead he wrote a pamphlet, *The Present State and Future Prospects of Mathematical and Physical Studies at the University of Oxford*, which was published in the Easter Term of 1832. In this pamphlet Powell argued that the mathematical and physical sciences form a proper part of a liberal education, and he concluded with a warning, urging the 'HIGHER CLASSES' to pay attention to physical and mathematical sciences or risk jeopardizing their social position:[5]

we cannot but derive from the circumstances of the *present age* the most powerful arguments for the necessity of increasing attention to the promotion of Physical and Mathematical Science. Scientific knowledge is rapidly spreading *among all classes* EXCEPT THE HIGHER, and the consequences must be, that that Class *will not long remain* THE HIGHER.

Unfortunately, this strident and often vituperative tone was counterproductive, particularly in Oxford.

The pamphlet also painted a bleak picture of the current situation in Oxford mathematics:[6]

out of the whole number of Candidates, though a certain portion had "got up" the four books of Euclid, not more than two or three could add Vulgar Fractions, or tell the cause of day and night, or the principle of a pump.

This was corroborated by another Examiner (Prof. Rogers), who recalled his experience of an oral examination in which

a student reproduced a proof from Euclid perfectly, except that in his diagrams he drew all his triangles as circles.

When challenged, the student replied:[7]

If you please, sir, I did not think we were required to learn the pictures.

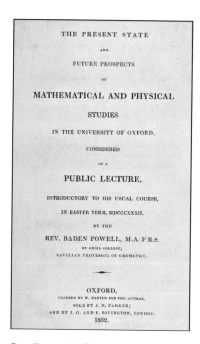 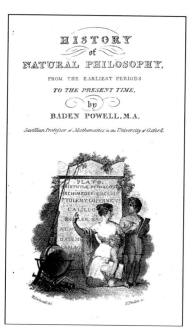

(*Left*) Baden Powell's pamphlet *The Present State and Future Prospects of Mathematical and Physical Studies in the University of Oxford*.
(*Right*) Powell wrote his *History of Natural Philosophy* to promote scientific awareness and remove public prejudices and misconceptions about the pursuit of science.

Although more a scientist than a mathematician, Powell was determined to fight his corner for mathematics, and he had numerous ideas for improving the standards and the appeal of the subject. Most of these were rejected by the University, but those reforms that were accepted had significant impact.

The first of these was to print and sell the mathematics examination papers, so as to provide problems on which students could practise and develop their skills, thereby gradually raising the standards; the first examination papers to be printed were taken in Easter Term 1828. Until 1918 there was an Easter Term as well as Michaelmas, Hilary, and Trinity Terms, and the Final Examinations could be taken in either Easter Term or Michaelmas Term. This was later changed to Trinity and Michaelmas Terms, and from 1883 they settled into Trinity Term alone.

The first examination question in Easter 1828 was

1. What decimal of a week is 1 hour 7 minutes and 14 seconds?

This presumably provided an opportunity for weak students to demonstrate that they could at least manage some elementary arithmetic. A similarly straightforward question appeared in the next examination later in the year:

8. A dealer sells a horse for 56l and gains as much percent as it cost. What did it cost?

There were also questions on simple trigonometry and algebra, but not all of the questions were so easy, as a question on Paper IX of the second examination illustrates:

2. Calculate a horizontal sundial for a latitude of 30°.

To answer this from first principles would require a good understanding of solid and spherical geometry, so the students may have learned techniques for doing it quickly or had tables to help them.

Baden Powell also raised money to endow a Mathematical Scholarship, awarded every year from 1831 on the basis of an examination, and tenable for three years. In 1844 this was converted into a Junior Scholarship and a Senior Scholarship, each tenable for two years; these provided an opportunity for the Scholars to get a start in the subject. During the professorships of Powell and Smith, these Scholarship holders included

- two future Savilian professors of astronomy: G. H. S. Johnson (1831) and W. F. Donkin (1837)
- two future Savilian professors of geometry: H. J. S. Smith (1851) and W. Esson (1860)
- a Sedleian Professor of Natural Philosophy: B. Price (1842)
- the first Waynflete Professor of Pure Mathematics: E. B. Elliott (1875)
- a future President of the Royal Society: W. Spottiswoode (1846).

William Donkin (1814–69) and Bartholomew Price (1818–98).

William Donkin and Bartholomew Price were later professorial colleagues of both Powell and Smith.

The examination paper for the Senior Mathematical Scholarship for 1851 (awarded to Henry Smith) has survived, and the questions provide a stark contrast with the 1828 Final Examinations, suggesting that Powell's ideas had worked. It includes questions on series expansions of differentiable functions, and the integration of an exponential function over a ball; both of these still appear regularly as examination questions, albeit now in the first year.

The foundation of the British Association for the Advancement of Science (commonly abbreviated as the 'British Association') provided another opportunity for Baden Powell, which he quickly seized.[8] With two of his scientific colleagues, Charles Daubeny and William Buckland, he arranged that during its first meeting in York in 1831 the new association should be invited to host the following meeting in Oxford. Powell and Buckland were unable to attend the York meeting, but Daubeny conveyed the invitation which was accepted, and the Association duly held its second meeting in Oxford in 1832.[9] This arrangement was mutually beneficial, as it gave the fledgling association the prestige

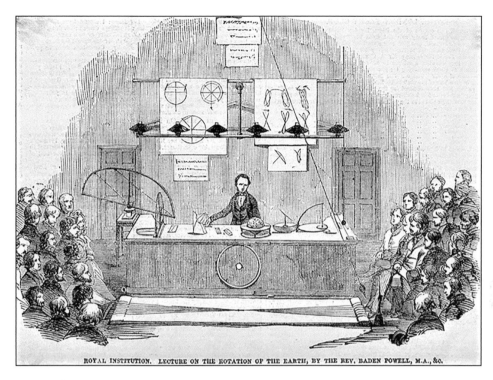

ROYAL INSTITUTION. LECTURE ON THE ROTATION OF THE EARTH, BY THE REV. BADEN POWELL, M.A., &c.

In May 1851 Baden Powell gave a public lecture on the rotation of the Earth at the Royal Institution in London.

of meeting at the oldest university in the country, while helping Oxford scientists to enhance their prestige within the University.

At this second meeting, Powell presented a 52-page *Report on the Present State of our Knowledge of Radiant Heat*.[10] During the meeting, four scientists (Sir David Brewster, Robert Brown, John Dalton, and Michael Faraday) received honorary doctorates from the University, even though the religious tests would have barred Dalton (a Quaker), Faraday (a Sandemanian), and perhaps Brewster (a Presbyterian) from studying at Oxford.

The British Association revisited Oxford in 1847, when it was attended by John Couch Adams from Cambridge and Urbain Le Verrier from Paris, who had independently calculated the orbit of the planet Neptune, subsequently discovered to great acclaim just a few months before the meeting. The highlight of the mathematical section's programme was a lecture by Adams on Neptune's discovery, in the presence of its co-discoverer.[11]

Around this time, Baden Powell started an extensive decade-long correspondence with the young George Gabriel Stokes at Cambridge.[12] Powell was really an experimentalist, and in 1847 he discovered that when he filled a hollow prism with liquid into which

(*Left*) The highlight of the 1847 British Association meeting in Oxford was a lecture by John Couch Adams on the recent discovery of Neptune.
(*Right*) At this meeting, Baden Powell presented a two-hour popular lecture on 'Falling stars' (shooting stars) to a large audience in the especially gas-lit Radcliffe Camera.

he inserted a transparent plate, the spectrum of transmitted light was interrupted by a series of dark bands. Stokes, who would become Lucasian Professor of Mathematics at Cambridge in 1849, was better acquainted with recent more powerful mathematical methods, and quickly started working on the theory. A collaboration ensued in which Stokes developed the theory and suggested new measurements, and Powell then made observations that confirmed Stokes's predictions.

(*Left*) Baden Powell's experiment in which he filled a hollow prism with liquid (M) and inserted a transparent plate (P). He discovered that the spectrum of transmitted light was interrupted by a series of dark bands.
(*Right*) George Gabriel Stokes (1819–1903).

Although Powell often had good ideas, he was handicapped by tactlessness and by his tendency to antagonize others by harping on about particular issues in discussion, and this probably damaged his effectiveness on the 1850 Royal Commission. As another member of this Commission, Archibald Campbell Tait (then Dean of Carlisle, and later Archbishop of Canterbury), explained in letters to his wife:[13]

Baden Powell [is] the most talkative of bores rendering business very difficult by his perpetual chatter.

Baden Powell remains at home writing a long harangue on Professorships which we have assigned to him to keep him quiet.

In general, Powell seems to have been most successful when he made common purpose with other science professors. One of the greatest achievements of Powell, Daubeny, and (especially) Henry Acland, the new professor of medicine, was to persuade the University to build a new centre and library for mathematics and the sciences, replacing the Radcliffe Camera which had previously housed the Bodleian Library's scientific books. By 1860 the new University Museum, although still not complete, was far enough advanced to host the British Association's third visit to Oxford – and, in particular, the famous debate on Darwin's recently published book, *On the Origin of Species*, at which the Bishop of Oxford, Samuel Wilberforce, clashed with T. H. Huxley and J. D. Hooker.[14]

Baden Powell was already a well-known proponent of evolutionary theories by the 1840s. He seems to have been led to his advocacy by theological considerations, believing that evolution provided so plausible a way for God to have accomplished the Creation that only atheists could fail to believe it, a suggestion that probably antagonized both sides of the debate.

When the work of Charles Darwin and Alfred Russell Wallace on evolution became public in 1858, it was welcomed by Baden Powell. In his essay, 'On the study of the evidences of Christianity', in a volume of *Essays and Reviews* (edited by Benjamin Jowett, Master of Balliol College, and published in February 1860), he wrote:[15]

a work has now appeared by a naturalist of the most acknowledged authority, Mr. Darwin's masterly volume on *The Origin of Species* by the law of 'natural selection'; – which now substantiates on undeniable grounds . . . the origination of new species by natural causes: a work which must soon bring about an entire revolution of opinion in favour of the grand principle of the self-evolving powers of nature.

Darwin returned the compliment in the third edition of his *Origin of Species* in 1861:[16]

Constructed in the late 1850s, the University Museum realized in brick and iron Oxford's mid-century aspirations to improve the facilities for teaching mathematics and science, in terms that recall the soaring confidence of medieval buildings such as the Divinity School.

The 'Philosophy of Creation' has been treated in a masterly manner by the Revd. Baden Powell, in his *Essays on the Unity of Worlds*, 1855. Nothing can be more striking than the manner in which he shows that the introduction of new species is 'a regular, not a casual phenomenon', or, as Sir John Herschel expresses it, 'a natural in contra-distinction to a miraculous process'.

Unfortunately, this tribute appeared too late for Powell, who had died on 11 June 1860, a few weeks before the evolution debate at the Oxford British Association meeting at which he had been due to speak. Not everyone was as complimentary as Darwin, however. Powell's 1860 contribution to *Essays and Reviews* proved highly controversial and would probably have brought him a Vice-Chancellor's summons had he not died before charges could be brought – instead, Jowett was summoned. Indeed, a later biography of Benjamin Jowett contained the following criticisms of Powell's essay:[17]

A more accurate title would have been *On the Evidential Value of Miracles*. It is an exasperatingly long-winded affair, abounding in italics which emphasize the writer's inability to clarify his thought, to extract what is to the point and to exclude what is not, and to express the result in readable language.

Another person who had been alarmed by the religious controversies in Oxford, and the furore stirred up by the *Essays and Reviews* in particular, was the mathematician George Boole. In a letter that he sent to Augustus De Morgan on 13 November 1860, he opined:[18]

I think [Jowett's] the best of the essays & for the sake of mathematics I am sorry to add Baden Powell's nearly the worst.

A man of strong Christian faith, Boole was worried by the doctrinaire approach taken by some at Oxford and so, when the Savilian professorship was advertised after Powell's death, Boole put in an application, but submitted no testimonials.[19] Meanwhile, the other Oxford mathematicians were petitioning the electors to select Balliol's Mathematical Lecturer, Henry Smith, and would not stand against him. With this show of local support and Boole's equivocation, Smith was duly elected Savilian Professor of Geometry. In 1861 he was elected to a Fellowship of the Royal Society.

Augustus De Morgan (1806–71) and George Boole (1815–64).

Henry Smith

The relevance of George Boole's work to computing has meant that he is now more widely known than Henry Smith, but Smith's mathematical work during his tenure of the Savilian chair achieved greater international recognition.

Henry John Stephen Smith was born on 2 November 1826 in Dublin and, despite spending most of his life in England, he was proud of his Irish roots. He won a scholarship to Balliol College, but before coming up in 1845 he contracted smallpox whilst visiting his family in Italy. During his first university vacation he went down with malaria, which forced him to spend two years convalescing on the Continent, and he put this interruption to good use by attending lectures at the Sorbonne in Paris.

(*Left*) Henry John Stephen Smith (1826–83).
(*Right*) A caricature of Henry Smith, illustrating the widespread affection for him across the University community. As a former student wrote after Smith's death: 'Alas . . . that his flowing beard and flying gown will be seen no more, for his place cannot be filled'.

In 1847 Smith was finally fit enough to return to Oxford. Two years later he graduated with First Class Honours in both Classics and Mathematics, having also won the top Classics Prize, to which he later added the Senior Mathematical Prize. The only other First Class performance in Mathematics in the Honours examination was that of Robert Faussett of Christ Church, who subsequently taught Charles Dodgson (better known as Lewis Carroll).

Later in 1849 Smith was elected as Mathematics Lecturer at Balliol College. He had some reservations as to whether he deserved this honour,[20] but Balliol had an ulterior motive. The University had just agreed to start the new Honour School of Natural Science, and Balliol decided to build the first college teaching laboratory to give practical instruction. Smith had attended lectures on chemistry in Paris, and Balliol wanted him to run their laboratory, as well as teaching mathematics. In preparation for this, he studied in the Old Ashmolean's basement laboratory with Nevil Story Maskelyne, the deputy professor of crystallography, and with August Hofmann at the Royal College of

Nevil Story Maskelyne's photographs of Baden Powell and Henry Smith, taken around 1850.

Chemistry in London.[21] It is to Maskelyne, and to his interest in photography, that we owe the first photographs of Baden Powell and Henry Smith, probably taken around 1850.[22]

Henry Smith's first student in the laboratory, Augustus Vernon Harcourt, had been impressed by his tutor, and he later worked with the mathematician William Esson in precise observations of chemical kinetics, anticipating some later work of the Swedish scientist Svante Arrhenius.[23] Harcourt and Esson studied what turned out to be an autocatalytic reaction, which greatly complicated their work, but in the course of this Esson noticed that the speed of reaction depended on the temperature in such a way that the reactions would stop at $-272.6°$C. Harcourt later instigated a collaboration between his student John Conroy and Stokes.[24]

Despite these other activities Smith did not neglect mathematics, and soon started to publish research papers. Number theory and geometry became his main interests, and his lectures on modern geometry were the first in England. His former pupil, John Wellesley Russell, later used them as the basis of his geometry books that were still in use in some English schools in the 1950s.[25]

Like Baden Powell, Henry Smith regularly attended the meetings of the British Association, and was soon commissioned by the Association to write a *Report on the Theory of Numbers*, an extensive survey of all the main branches and methods of number theory from the Greeks onwards. This appeared in six parts in the Association's proceedings, starting in 1859 and finishing in 1865, some 236 pages in all, and was praised by Leopold Kronecker, professor of mathematics in Berlin, for its lucidity and insight.

Within Oxford, Smith joined three other college lecturers in an important reform of the teaching. Traditionally, each mathematics lecturer would just teach students from his own college and would cover the entire syllabus, but around 1867 Henry Smith, William Esson (Merton College), Charles Faulkner (University College), and Charles Price (Exeter College) introduced the Combined College Lecture scheme; Price had been one

---

Report on the Theory of Numbers.—Part I.
By H. J. STEPHEN SMITH, M.A., *Fellow of Balliol College, Oxford.*

1. THE 'Disquisitiones Arithmeticæ' of Karl Friedrich Gauss (Lipsiæ, 1801) and the 'Théorie des Nombres' of Adrien Marie Legendre (Paris, 1830, ed. 3) are still the classical works on the Theory of Numbers. Nevertheless, the actual state of this part of mathematical analysis is but

Henry Smith's *Report on the Theory of Numbers* for the British Association.

of Smith's early pupils at Balliol. Each lecturer now taught about one quarter of the syllabus to students from the four participating colleges, and other students could attend on payment of a fee. This scheme proved successful, and by 1878 it had become a prototype for teaching across all subjects.

It is sometimes thought that this inter-collegiate scheme was first mooted by Esson, but the important part played by Smith is testified by the fact that the scheme was suspended for two years following his death. Ironically, Esson was the only member of the scheme who got into trouble with his college for failing to obtain prior approval for the scheme. Later, as we see in Chapter 5, Esson deputized for James Joseph Sylvester, Smith's successor as Savilian Professor of Geometry, and was eventually elected as Sylvester's successor in that position.

Henry Smith's personality was very different from Baden Powell's, and in debates he was renowned for his ability to 'say the happy word which quelled a rising storm'. Indeed, Benjamin Jowett, who did not mince his words and had thought Powell 'odious', called Smith 'the most gracious man I ever knew'.[26]

Smith's wide experience as a prize-winning graduate in Classics who had run a science laboratory, and who was gaining a growing reputation within mathematics, earned him the respect from all parts of the University which Powell had never attained. Indeed, as John Conington, the Professor of Latin, once remarked:[27]

I do not know what Henry Smith may be at the subjects of which he professes to know something, I never go to him on a matter of scholarship in a line where he professes to know nothing, without learning more from him than I can get from anyone else.

Smith was therefore a natural choice for University committees such as Oxford's Hebdomadal Council, which often entrusted him to introduce its proposals when they were debated in Congregation (the assembly of resident academics). Outside Oxford, Smith became the first President of the Meteorological Council and a governor of various schools and of Bristol University.

Perhaps Smith's most important appointments were from 1870 to 1875 to the Royal Commission (Devonshire Commission) on Scientific Instruction and the Advancement of Science, and from 1877 to 1878 to the Selborne Commission on Oxford University. The latter undertook important reforms to the University, recommending that

- the combined lecture scheme should be extended
- three new professorships should be established
- all endowed professorships should be attached to particular colleges.

These were all achieved. The combined lecture scheme became the CUF (Common University Fund) lectureship arrangement across all faculties, new professorships included the Waynflete chair of pure mathematics, and the Savilian chairs were attached to New College with which they had enjoyed a long association.

Some of Smith's most important mathematical contacts were from outside Oxford. William Spottiswoode had overlapped with Smith as a student at Balliol. He joined the family business, Eyre and Spottiswoode, Printers for Her Majesty's Stationery Office, but that left him sufficient time to research in mathematics and to run his own physical laboratory. He was elected to a Fellowship of the Royal Society in 1853, and in 1878 became its president, an office that he held until he died from typhoid in 1883, just a few months after the death of Henry Smith.

James Glaisher studied at Cambridge University, where he became a Fellow at Trinity College, and was elected to a Fellowship of the Royal Society in 1875. He had known Smith even before going to Cambridge, and the two met regularly. It was Glaisher who persuaded Smith to publish papers based on some of his British Association talks in the *Messenger of Mathematics* during his time as editor. The two later worked together on elliptic functions and, at Glaisher's suggestion, Smith wrote a *Memoir on the Theta and Omega Functions* that provided the background for Glaisher's tables of these functions.

Spottiswoode, Smith, and Glaisher became early members of the London Mathematical Society soon after its foundation in 1865, and all three served as its president: William Spottiswoode from 1870 to 1872, Henry Smith from 1874 to 1876, and James Glaisher from 1884 to 1886.

In 1869 Henry Smith shared the Berlin Academy's Steiner Prize with Hermann Kortum of Bonn for their work on a geometrical problem that had been posed by the Academy. Whereas Smith's collected mathematical papers fill two large volumes, Kortum published only two papers: his doctoral thesis and his solution to this prize problem. Reminiscences of him in later life suggest that he was not really interested in mathematics,[28] and he died of a chill that he caught whilst sailing as a guest on the Kaiser's yacht.

In 1874 Henry Smith was appointed Keeper of the University Museum, following the death of John Phillips, the professor of geology; Smith's sister, Eleanor, remarked in a letter of 29 October 1874 that he accepted 'for the honour and to be near the croquet ground'.[29] Henry and his sister, who had shared accommodation since the death of their mother, moved into the Keeper's House; here he had his office in the servants' quarters on the top floor because he liked the view across to Shotover from the dormer window. There is a bust of Smith in the University Museum.

In the same year, Smith was elected to a professorial fellowship at Corpus Christi College. Until that time he had continued as Mathematical Lecturer at Balliol, combining his

William Spottiswoode (1825–83) and James Glaisher (1848–1928).

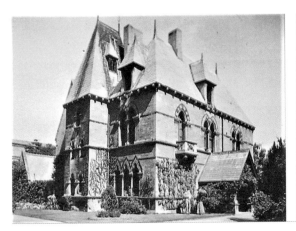 

(*Left*) The University Museum's Keeper's House.
(*Right*) Henry Smith's bust in the University Museum.

Charles Hermite (1822–1901), Felix Klein (1849–1925), and Ferdinand von Lindemann (1852–1939).

duties there with those of the Savilian chair. Balliol, however, was reluctant to lose him completely and elected him as its second honorary fellow (the poet Robert Browning being the first). The new arrangements seemed to leave Smith more time, and in the next couple of years he published many papers, several at Glaisher's suggestion.

By this time Henry Smith was hosting European mathematicians on visits to Oxford. In 1873 he became President of the Mathematical Section of the British Association, and he invited Charles Hermite from Paris, and Felix Klein who was about to move to Erlangen in Bavaria. Hermite talked on a proof that the exponential number e is irrational, presumably as an easier window on his proof that it is transcendental.[30] Smith had already known Hermite since at least April 1865, when the two of them had dined with Hermann von Helmholtz in Paris,[31] and it is likely that he also played a part in Hermite's election to a Fellowship of the Royal Society in that year. (Klein was younger, and was elected to a Royal Society Fellowship a decade later.) In 1876 Ferdinand von Lindemann from Würzburg visited Smith in Oxford, and it is reported that they discussed the possibility of proving that π is transcendental, an achievement that Lindemann published in 1882.

Smith's mathematical work

Whilst preparing his *Report on the Theory of Numbers*, Henry Smith was engaged in researches of his own. Unlike Powell's work, many of Smith's papers investigate problems that are as easy to state as they are difficult to solve. In this section we present a range of topics to which Smith made noteworthy contributions.

Linear Diophantine equations

Euclid's *Elements*, often thought of as mainly a textbook on geometry, also includes three books on arithmetic. This area fascinated Greek mathematicians, and Diophantus of Alexandria also wrote extensively on the subject, studying 'Diophantine equations', in which all the featured numbers are integers (whole numbers), as are the solutions we seek. For example, consider the equations $2x = 4$ and $2x = 5$. Only the first of these Diophantine equations has an acceptable integer solution, $x = 2$; the second does not, because $x = {}^5/_2$ is not an integer.

Such equations had been studied for two millennia, and many methods were known for discerning when equations have solutions and finding them when they exist. But it was Henry Smith, in a paper published in the *Philosophical Transactions of the Royal Society* in 1861, who completely solved the problem of finding when linear Diophantine equations have a solution and obtaining one when it exists. ('Linear' means that the equations contain no higher powers or products of x and y, such as x^2, y^3, or xy.)

Smith was also the first to give necessary and sufficient conditions for simultaneous equations to have real solutions, a result that was first published in *An Elementary Treatise on Determinants* by Charles Dodgson.[32] At school one learns how to solve simultaneous linear equations, such as

$$2x + 3y = 7, \quad x + 4y = 6,$$

by elimination. For example, subtracting the first equation from twice the second gives $5y = 5$, so that $y = 1$: if we had replaced the first equation by $2x + 3y = 3$, then we would have obtained $5y = 9$, with no integer solution. Then, substituting $y = 1$ into the second equation we obtain $x = 2$. Alternatively, we can find x by subtracting three times the second equation from four times the first, to obtain $5x = 10$. It is no coincidence that the multiple 5 occurs both times; it is known as the *determinant* of the coefficients.

Smith gave a method for rewriting any system of linear simultaneous equations with integer coefficients in any number of unknowns x, y, ...; to do so, he introduced new unknowns X, Y, ... (given explicitly in terms of x, y, ...), and added and subtracted multiples of equations as above to produce what is now known as the *Smith normal form* of the equations

$$AX = P, \ BY = Q, \dots,$$

where A, B, ... and P, Q, ... are integers and each coefficient A, B, ... divides all subsequent coefficients. In this form, the equations can be solved precisely when A divides P,

B divides Q, and so on. (In the above example we could take $X = x + 4y$ and $Y = y$ to rewrite the equations as $X = 6$ and $5Y = 5$, with $A = 1$ and $B = 5$; then $y = Y = 1$ and $x = X - 4y = 2$.) There are many choices for X, Y, ..., but they all give the same values for A, B, ... with the above division property.

Smith's study of linear Diophantine equations had exploited the Cambridge mathematician Arthur Cayley's new theory of matrices; this had been published only in 1858, but Smith must have heard about it earlier and was already corresponding with him in 1857. Later, on 7 June 1864, Smith and Dodgson enjoyed a breakfast meeting with Cayley, who was visiting Oxford University to receive an honorary Doctorate in Civil Laws.[33]

Sums of squares

In 1621, while preparing his edition of Diophantus's *Arithmetica*, the French mathematician Claude Gaspard Bachet de Méziriac conjectured that every positive integer can be written as the sum of four perfect squares. This conjecture was studied by Pierre Fermat and Leonhard Euler, among others, but the final proof was found by Joseph-Louis Lagrange in 1770. As examples we note that

$$2 = 1^2 + 1^2 + 0^2 + 0^2 \text{ and } 7 = 2^2 + 1^2 + 1^2 + 1^2;$$

the second of these shows that we cannot manage with fewer than four squares. Writing

$$4 = 1^2 + 1^2 + 1^2 + 1^2 = 2^2 + 0^2 + 0^2 + 0^2$$

shows that there can be more than one way of representing a number as a sum of four squares, and raises the general question:

In how many ways can one express a given positive integer as the sum of a chosen number of squares?

Smith had already published an elegant two-page paper on sums of two squares in the prestigious German *Crelle's Journal*; unusually his paper was in Latin, possibly because Smith found it easier to write in that language than in German, or because he intended it as a homage to *Disquisitiones Arithmeticae*, a treatise on number theory that had been written in 1801 by the great German mathematician Carl Friedrich Gauss, which was also in Latin.

Adrien-Marie Legendre and Gauss had both dealt with sums of three squares, and Carl Gustav Jacob Jacobi had found the number of ways of expressing a positive integer as the sum of four squares, or of six squares, as a by-product of his work on elliptic functions. This left the two cases of five and seven squares, tackled by Gotthold Eisenstein in the 1840s, but Eisenstein's early death meant that his investigations covered only special cases.

In an 1867 paper in the *Proceedings of the Royal Society*, Smith produced the answers in these cases. Moreover, he developed a uniform method that worked for the other known cases as well, and which exploited his earlier work on linear Diophantine equations.

Some applications

In two dimensions the points with integer coordinates (with respect to horizontal and vertical axes) form a lattice. Smith showed that in eight dimensions there is a similar lattice in which all eight coordinates are integers, and also that there is a less obvious lattice that cannot be obtained from this one by rotations, squeezing, shearing, or other such tricks. This new lattice was soon constructed explicitly and has found many interesting applications. In mathematics it is associated with interesting exceptional symmetries known as E_8. A more recent application was presented in 2016 in an important paper of Maryna Viazovska, which showed that the closest packing of spheres in eight dimensions occurs when the centres of these spheres appear at the points of Smith's lattice.[34]

Smith's own work was mainly in pure mathematics, and while summarizing a lecture he once joked:[35]

It is the peculiar beauty of this method, gentlemen, and the one which endears it to the really scientific mind, that under no circumstances can it be of the smallest possible utility.

But Smith by no means disparaged applications. He was one of the first mathematicians to draw attention to James Clerk Maxwell's theory of electromagnetism, and he collaborated with Maskelyne in work on crystal structures. In 2010 the exceptional symmetries associated with his lattice were detected in cobalt niobium crystals at very low temperatures.[36]

Fractal sets

But this is not the only recent unexpected application of Smith's ideas. In 1875 he published a paper 'On the integration of discontinuous functions', a somewhat unusual topic for him, being neither on number theory nor on geometry. He had been giving a lecture course on integration around this time, so he may have been led to the subject by that.

A graph may be continuous – in lay terms, we can draw it without lifting our pencil off the paper – or it may be discontinuous, so that there are some 'jumps'. Integration provides a way of calculating 'the area under the graph'. This certainly makes sense for continuous graphs, and also for those with only a finite number of jumps, as the area can be found by adding the areas under the separate sections between the jumps. Smith investigated the boundary between those graphs with infinitely many jumps for which the area can be found, and those graphs for which it cannot.

In 1875 Henry Smith's investigations led him to the following construction. Choose a number n (> 2), divide a given interval into n parts, and remove the last of these; then divide each of the remaining subintervals into n parts, and continue the process for ever; for example, when $n = 3$, all of the points in the right-hand half of the interval are eventually removed, while those remaining in the left-hand half form what is now known as 'Cantor's ternary set' – the German mathematician Georg Cantor had rediscovered it in 1883, eight years after its discovery by Smith. It is now recognized as a *fractal set*, and Smith was probably the first to construct such a set.

Smith showed that the area under a graph with jumps at the surviving points can be found, but if one modifies the process so that the number of subdivisions of the interval at each stage is three times that at the previous stage (first subdivide it into 3, then 9, then 27, etc.), then the corresponding area cannot be found. Had Smith's discovery received more attention by Continental mathematicians, they might well have speeded the systematic development of integration theory that took place at the end of the century.

In 2018 Howard Colquhoun, Ricardo Grau-Crespo, and their collaborators found that, when the intervals are divided into four parts ($n = 4$), Smith's construction enabled them to understand the nuclear magnetic resonance (NMR) spectrum of what is now called a Howard molecular copolymer complex.[37] Screening effects within the Howard

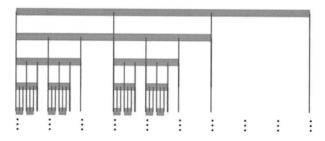

Smith's construction with $n = 3$; the shaded points that remain form what is now known as 'Cantor's ternary set'.

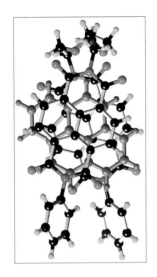

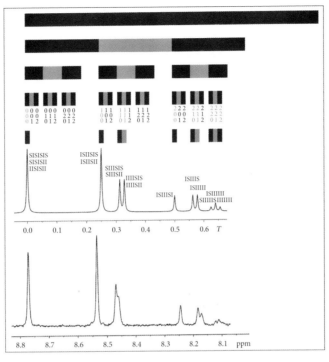

(*Left*) The X-ray structure of a small-molecule model for the copolymer complex whose NMR spectrum can be described by a Smith fractal set with $n = 4$.

(*Right*) Smith's quaternary set construction, the NMR simulation based on that set, and the measured NMR spectrum. Numerical simulations after subintervals have been discarded appear in the upper graph and closely match the experimental observations in the lower one.

complex result in only a few of the subintervals remaining after three subdivisions influencing the spectrum. Given his training in chemistry, Smith would probably have been fascinated by this application of his work on integration, almost 150 years after its publication.

The Paris Grand Prix

In 1881 Henry Smith had a fall, and while recuperating he noticed that the latest prize problem advertised by the French Academy of Sciences concerned the decomposition of numbers as sums of five squares. Somewhat perplexed by this, Smith wrote to Glaisher on 17 February 1882, asking his advice:[38]

The Paris Academy have set for their Grand Prix this year the theory of the decomposition of numbers into five squares . . . In the Royal Society Proceedings vol xvi, pp 207–208 I have given the complete theorems not only for five, but also for seven squares . . . Ought I to do anything on the matter? My first impression is that I ought to write to Hermite, and call his attention to it. A line or two of advice would really oblige me as I am somewhat troubled and a little annoyed.

Glaisher confirmed that he should indeed write to Hermite, which Smith did on 22 February. Four days later Hermite replied,[39] expressing his embarrassment that the problem must have been suggested by a member of the Academy who was unacquainted with Smith's work, but hoping that Smith would help the Academy to save face

and consider whether it is convenient for you to respond to the appeal of the Academy to those who love Arithmetic; in any case rest assured that I shall see that the Commission is aware of your works, when it reports to the Academy on the memoirs submitted.

In other words, Smith should rewrite his paper in French and submit it for the Prize.

Despite illness, Smith managed to do this, but on 9 February 1883 he died, two months before the prizewinners were announced. Smith's death was met with dismay, both among mathematicians and within Oxford; the reaction of Mark Pattison, Rector of Lincoln College, gives an idea of Smith's esteem within the University:[40]

The most accomplished man in the whole university, at once scientific and literary, 20 MAs – any 20 – might have been taken without making such a gap in the mind of the place.

On 1 April 1883 the prize was awarded jointly to Henry Smith and to Hermann Minkowski, an 18-year-old student at the University of Königsberg in East Prussia. Hermite had apparently failed to alert the Academy to Smith's priority, and when Eleanor Smith wrote to him to remind him of his promise, he replied that it was just an absolutely unintentional lapse of memory.[41]

It did not take journalists long to discover from Henry Smith's obituaries that the Academy had posed a problem that he had already solved fifteen years earlier. Coupled with the fact that the Academy had noted the similarity between the approaches in the two prizewinning solutions, and perhaps also reflecting continuing French resentment of Prussians after the 1871 Franco–Prussian war, the obvious conclusion was that Minkowski's prize entry had been plagiarized. After weeks of continued criticism, the Academy tried to bring the matter to a close by awarding full prizes to both Minkowski and Smith, rather than their having to share it.

We now know that Minkowski went on to a distinguished career in mathematics, and might well have solved the problem himself – or, if he had known Smith's 1867 paper,

Henry Smith, in old age, with Hermann Minkowski (1864–1909).

would have assumed that the Academy knew it also, and that they were just trying to elicit further details.

As a postscript to the story, we note that on 5 January 1900, responding to his friend David Hilbert's request for possible ideas for his address to the Paris International Congress of Mathematicians later in that year, Minkowski wrote that the only thing which came to mind was Smith's presidential address to the London Mathematical Society, *On the Present State and Prospects of Some Branches of Mathematics*, in which he discussed interesting questions and new ideas for mathematicians to study.[42] We do not know whether this had any influence on Hilbert's now famous lecture describing a series of problems that merited investigation by mathematicians in the coming century, but it certainly suggests the respect in which Minkowski, whose criticism of other mathematicians was often scathing, held Henry Smith.

James Joseph Sylvester, shortly after his arrival in Oxford in 1884. This image illustrated an article on him in a series of 'Scientific Worthies' that was published in *Nature* in 1889.

CHAPTER 5

James Joseph Sylvester

KAREN HUNGER PARSHALL

Following the tenure of Henry Smith, James Joseph Sylvester served as Savilian Professor of Geometry from 1883 to 1894. The first Jew to hold an Oxbridge chair, Sylvester came to Oxford from the inaugural professorship of mathematics at The Johns Hopkins University in Baltimore, Maryland, USA. While mainly focusing on his time in Oxford, this chapter contrasts the situations that Sylvester encountered at Oxford and Hopkins at the end of the 19th century.

James Joseph Sylvester received word of the sad news of Henry Smith's untimely death in a letter of 12 February 1883 from his longtime friend Arthur Cayley, Cambridge's Sadleirian Professor of Pure Mathematics. That letter had to cross the Atlantic to reach Sylvester for he had not been employed in England since 1870. He had, in fact, left his homeland in 1876 to accept the inaugural professorship of mathematics at The Johns Hopkins University in Baltimore, Maryland, and had spent the seven intervening years pursuing his invariant-theoretic researches, animating the first research-level programme in mathematics in the United States, and frequently wishing he were not so far from home, despite the many American friends he had made.

Cayley saw in Smith's tragedy an opportunity for the still vigorous 68-year-old Sylvester to return home to assume the sort of position that his mathematical achievements merited. After all, with Cayley, he had co-founded a British school of the algebraic theory of invariants in the 1850s that had earned him an international reputation. That,

Karen Hunger Parshall, *James Joseph Sylvester*. In: *Oxford's Savilian Professors of Geometry*.
Edited by Robin Wilson, Oxford University Press. © Oxford University Press (2022). DOI: 10.1093/oso/9780198869030.003.0005

however, had never been enough to overcome the handicap of his Jewish heritage in English academe and to secure for him an Oxbridge position. Sylvester's journey to the moment he received the news of Henry Smith's death – with its possibility of the assumption of Oxford's Savilian professorship of geometry – had had many twists and turns.

A difficult professional road

Sylvester was born in London on 3 September 1814, the son of Miriam Joseph and her husband, the well-to-do Liverpool silversmith-turned-London entrepreneur Abraham Joseph.[1] As a boy, James attended a private boarding school in Highgate that, in catering to the Anglo-Jewish elite, provided its sons with both a religious (Hebrew and biblical studies) and secular (Latin, Greek, and mathematics) education. So precocious was the young James in his mathematical work that his teacher arranged for him to be examined in algebra by Olinthus Gregory, the professor of mathematics at the Royal Military Academy, who 'pronounced him to be possessed of great talents' and who continued to follow the boy's progress thereafter.[2] This prompted James's move, first to a school in Islington and then, in 1828, to the non-sectarian London University (after 1836 called University College, London (UCL)). At the latter, he enrolled at age 14 under the name of James Joseph Sylvester – following in the footsteps of his elder brothers in taking the surname 'Sylvester' – and was a student under one of England's best mathematicians of the day, Augustus De Morgan, in the new university's first entering class.

Sylvester's student association with De Morgan and his university was, however, short-lived. In the wake of an incident in the refectory in which Sylvester allegedly took a table knife 'with the intention of sticking it into a fellow student who had incurred his displeasure', his family withdrew him, owing to 'extreme youth' and to 'the fact of his requiring constant control and attention'.[3] He was next sent to Liverpool, where he lived with his elder sisters and attended the Royal Institution School. Although an academic success, Sylvester's time in Liverpool brought more social troubles for a young man who did not suffer fools gladly and who was sensitive to 'a good many schoolboy remarks respecting his creed'.[4]

Sylvester's next battleground centred, between 1831 and 1837, on a Cambridge University in the throes of a debate that would have removed various restrictions on those who did not adhere to the teachings of the Church of England – in particular, the requirement that, in order to take his degree, a student had to subscribe to the Church's

Thirty-Nine Articles of faith.[5] Although not a practising Jew, Sylvester, unlike some other non-Anglicans, was unwilling to swear to something in which he did not believe. As a result, although placed Second Wrangler in the Mathematical Tripos in 1837, Sylvester could not actually receive the degree that he had earned with distinction.

With Oxbridge closed to him as a career path, Sylvester faced the question: 'What next?'. Fortunately, an answer presented itself in October 1837 when he learned that UCL's chair of natural philosophy was vacant. He applied and learned in November that he had received the nod.[6] Maybe there would be a place for him in English academe after all.

Unfortunately, Sylvester's second sojourn at University College proved no more satisfying than the first. Over the course of his three-and-a-half years on the faculty, his enrolments, and hence the fees on which his salary depended, were low, and he found the experimental side of his position – setting up and conducting laboratory demonstrations in a dank and leaky room – thoroughly uncongenial. What he did find rewarding was the opportunity to pursue mathematical researches of both an applied and a pure cast, and to discuss the fruits of his labours at meetings of the Royal Society of London to which he was elected a Fellow in 1839 at the young age of 24. Sylvester also took the opportunity to acquaint himself better with work being done on the other side of the English Channel by attending meetings of the Académie des Sciences in Paris. Even at this early point in his career, Sylvester prized international connections. He would continue to make and to nurture them throughout his lifetime.

An international opportunity of a different sort presented itself in 1841. Charles Bonnycastle, an Englishman transplanted to the non-sectarian University of Virginia in the United States, had died in October 1840, thereby creating an opening in the chair of mathematics.[7] Two of Sylvester's older brothers had already emigrated to the United States and were prospering in New York City. A move to Virginia and an actual professorship in mathematics, rather than of natural philosophy, might be just the answer for an aspiring mathematician limited in English academe by his Jewish heritage. Sylvester put together an impressive set of testimonials, made the appropriate contacts with the University's agents in London, and even did what was necessary to take a Bachelor's degree, on the basis of his work at Cambridge, as well as a Master's degree from Trinity College, Dublin, where religious tests for degree candidates had been abolished in 1794. It was thus a fully certificated 27-year-old Sylvester who accepted Virginia's offer and made a bold transatlantic move to Charlottesville in 1841.

If only the University that he encountered had been the congenial institution of higher education of which he had hoped to be a part. Although he was ostensibly welcomed

J. J. Sylvester, c.1840, in a painting by George Patton.

on his arrival in November, all was not as it appeared. The students were rowdy and disrespectful; the broader community in the slave-holding South exhibited strains of xenophobia, as well as religious prejudices against non-Protestants. Sylvester lasted just four-and-a-half months. In the wake of a physical encounter with a student whom he had had to discipline repeatedly in class, he precipitously tendered his resignation and left Charlottesville to take refuge with his brothers in New York.

For the next year and a half, Sylvester struggled, as he put it, 'against an adverce tide of affairs', among them a failed proposal of marriage and repeated failures to secure a new academic position.[8] He *was* successful, however, in making scientific contacts in the American Northeast, particularly with the Harvard professor of mathematics Benjamin Peirce and with the Princeton physicist Joseph Henry. Both would remain his friends until their deaths in 1880 and 1878, respectively. It was nevertheless a dejected Sylvester who returned to London at the end of 1843 with no job prospects on the horizon.

As he cast about for what he might do, Sylvester rekindled his mathematical research, re-established his scientific acquaintances with De Morgan and others, made new contacts with such influential figures as Lord (Henry) Brougham, and participated in September 1844 in the meeting of the British Association for the Advancement of Science in York.[9] By December he had accepted the position of actuary at the Equity and

'The Student', an 1853 sketch of a University of Virginia student that captures the arrogant attitude characteristic of the day.

Law Life Assurance Society in London that he would hold until 1855. As a 'professional man', he earned a respectable salary of £250 per annum and lived in back rooms in the firm's office at 26 Lincoln's Inn Fields.

The ten years from 1845 to 1855 found Sylvester studying for the bar at the Inner Temple in order to qualify himself for an eventual stockholding directorship of the Equity and Law Life, writing up policies for his firm's clients, engaging actively in the new Institute of Actuaries (founded in London in 1848), being called to the bar in 1850, and trying to keep his hand in with his mathematical research. They also found him, after 1847, developing the deep friendship and mathematical give-and-take with fellow mathematician-turned-conveyancing lawyer Arthur Cayley that would characterize the rest of their lives.[10] By the early 1850s, almost daily mathematical exchanges between them had resulted in the formation of a new area of mathematics.

Their algebraic theory of invariants involved the behaviour of forms – homogeneous polynomials of degree n in m variables – under linear transformation of those variables.[11] To take a simple example ($n = m = 2$), consider what Sylvester and Cayley termed a 'binary quadratic form',

$$Q = ax^2 + 2bxy + cy^2,$$

Arthur Cayley (1821–95).

where a, b, c are real numbers, and a linear transformation T of its variables with non-zero determinant r that takes Q to

$$R = Ax^2 + 2Bxy + Cy^2.$$

The 'discriminant' $b^2 - ac$ in the coefficients of Q is an invariant, because

$$B^2 - AC = r^2(b^2 - ac)$$

– that is, the discriminant is changed under linear transformation only up to a power of the determinant. The question then became to find all of the invariants for a given form. This would occupy Sylvester and Cayley for the rest of their active mathematical lives, at the same time that their work toward its solution would establish their mathematical reputations internationally.

They continued to work from their respective rooms in Lincoln's Inn until 1855, when Sylvester succeeded (following a failed try a year earlier) in being named professor of mathematics at the Royal Military Academy in Woolwich in south-east London.[12] He would once again be in academe, where, he told his political benefactor Lord Brougham, he hoped 'to consecrate my life and powers to the study & humble imitation of the spirit of the great masters of our science, to serve and if possible advance which is in my mind one of the noblest objects of ambition which one can propose to oneself'.[13]

The fifteen years that he spent at Woolwich found him producing prodigious amounts of new research, successfully editing the *Quarterly Journal of Pure and Applied Mathematics*, and garnering accolades, such as his election as Foreign Corresponding Member of the Académie des Sciences, the Presidency of the London Mathematical Society from 1866 to 1868, and the Presidency in 1869 of Section A (Mathematics and Physics) of the British Association for the Advancement of Science. His Woolwich years ended, however, in 1870, when changes in the regulations forced him to retire without a pension at age 55.

Six years followed during which he successfully fought the military authorities for his rightful pension, spent much time at his London club, the Athenaeum, participated in the working men's educational movement, lectured at the Royal Institution and elsewhere, journeyed to Europe to take part in its scientific culture, and indulged the poetic muse who (he believed) graced him.

These years did not, however, see much in the way of new mathematical research from his pen.[14] Still energetic and vital, he longed to be back in the mathematical fray and seriously contemplated throwing his hat in the ring for the vacant chair of mathematics at the University of Melbourne in Australia. Although he credited Cayley for saving him 'from the Antipodes', his friend fully supported him in his ultimately successful bid as the inaugural professor of mathematics at The Johns Hopkins University, a new educational experiment located in Baltimore, Maryland.[15] At the age of 61, Sylvester would make another transatlantic move, this time to animate the first research-level programme in mathematics in the United States.

After three largely unfulfilling professorial attempts, the fourth was a charm for Sylvester.[16] His new position required him to pursue his own researches and to build and foster a graduate programme in mathematics. It also provided him with the means to launch, in 1878, and edit the research-oriented *American Journal of Mathematics*. Freed from undergraduate teaching, he developed an idiosyncratic but effective style of graduate teaching that engaged his students in his then-current research, by challenging them first to immerse themselves in his ideas and then to prove their own theorems. He also worked under a President, Daniel Coit Gilman, who fully supported him, who accepted his foibles, and who became his friend. Still, after seven-and-a-half years, Sylvester was

The Mathematical Seminary at The Johns Hopkins University, c.1890. The plaster models of surfaces on the shelves were by Ludwig Brill of Germany.

homesick and longed to return to England. This was when a fifth and final professorial opportunity, almost beyond his wildest imagination, presented itself: the Savilian chair of geometry in Oxford.

A chance at Oxford?

When he occupied UCL's professorship of natural philosophy late in 1837, Sylvester, as a Jew, would have been ineligible for fellowships – much less professorships – at England's ancient universities.[17] The spirit of reform that had been afoot in the Cambridge of his student days to remove disabilities for Jews had, from the mid-1850s to the early 1870s, finally brought many of the desired changes. In the wake of this new legislation, Sylvester's *alma mater*, Cambridge University, had already acknowledged his student and subsequent achievements in 1872 by awarding him his long overdue BA and MA degrees, *honoris causa*. In June 1880 his former college, St John's, had acknowledged him even more completely – as a member of its actual fellowship – when they elected him an honorary fellow. Perhaps not to be outdone, Oxford awarded him an honorary Doctorate of Civil Law, also in June 1880. As these changes imply, in 1883 it was finally possible for Sylvester, as a professing though not a practising Jew, to become Savilian Professor of Geometry in succession to Henry Smith, but did he dare push the point?

These and other thoughts weighed on him. By March 1883 he was close to a decision: 'I shall most probably offer myself as a Candidate for the succession to his professorship', he confided to Cayley. 'Do you think I am likely to be appointed?'[18] He laid out some of the pros and cons. 'If the chances are considerably against me', he argued, 'it would be impolitic to offer'. Still, he continued, 'perhaps even if impolitic it would be right on my part to do so by way of testing what I consider – although you may not perhaps agree with me – an important principle' – namely, the religious freedom guaranteed by the Universities Tests Act of 1871 and Oxford's willingness to open its faculty to Jews. By April, he had made his decision and had submitted his formal application for the chair, through his friend and one of the members of the selection committee, William Spottiswoode.[19]

Smith's successor was supposed to have been named no later than 7 July, but the surprise of Spottiswoode's death late in June resulted in the election's postponement and much anxiety for Sylvester. It was finally announced that a decision would be made on 4 December 1883. Sylvester would just have to wait, although patience was not one of his virtues. The vacancy, however, had prompted him to make a major life decision. As he told Cayley, he had tendered his resignation at Johns Hopkins, effective from 1 January 1884, regardless of the outcome in Oxford. 'If I am not elected', he confided, 'my present purpose is either to buy a life annuity with my savings of the last 7 years or to go in for some business partnership'.[20] Those plans were mooted when Sylvester learned by telegram from Cayley on 5 December that he had been elected. He would be the first professing Jew to hold a chair at Oxbridge.[21]

The prodigal son returns

The Oxford that Sylvester entered was in the throes of a difficult transition, which had (in some sense) begun in 1850 with the Royal Oxford University Commission, from a collection of essentially ecclesiastical colleges into a modern university. Other commissions had followed – in particular, one in 1877 that had recommended the creation of specialized 'Boards of Faculties', the members of which would give lectures to, and examine, any and all interested students, regardless of their college affiliation. This prompted agitation on the part of the college tutors, who worried that their position might be undermined by these more specialized boards. They ultimately maintained their power within the University when it was decided that the professors would sit as *ex officio* members of the boards, but only the college tutors would have voting rights to elect new members.[22] The

Oxford that Sylvester encountered in 1884 was one that was largely under the sway of these 1877 reforms.

New College, in which he lived, was then dominated by the Warden, James Sewell. Dubbed 'The Shirt', Sewell was a man of the old school who had come to New College as a student in 1827 and, except for a period of a few months in the 1830s, had lived in College ever since. His mindset was very much that of the *status quo*: he saw no need to expand the curriculum or the student body; he saw no need for specialization or an academic profession; he had little thirst personally for academic attainments. Colleagues, however, like Hereford George, a tutor and a modern who viewed college teaching and research as a viable career option, increasingly lobbied for change in the late 1870s and early 1880s. Sewell did not officially oppose or attempt to block their efforts.

Sylvester's first months as Savilian professor found him between two worlds, unclear as to how or whether he would fit in. In response to congratulations on his new post from the Master of St John's, Cambridge, for example, Sylvester allowed that 'I . . . find every body friendly and disposed to make me feel at home in my new surroundings', although he also wondered whether[23]

I may perhaps be able to serve some useful purpose as a common term (to speak in Algebraical phrase) between our two great Universities and the Universities with which I have been affiliated on the other side of the Atlantic.

Gilman had consciously encouraged and benefited from Sylvester as a mathematical link between Baltimore, England, and the Continent. In seeking to define his new niche, Sylvester thought that perhaps he might still play such a role as the new Savilian professor.

One predetermined aspect of that niche, though, was teaching – but it was not the kind of teaching to which he had become accustomed, and in which he had excelled at Hopkins. Rather, it was lower-level undergraduate instruction in the prescribed subject of geometry, never one of his favourite topics and one that he had systematically striven to eliminate from his *algebraic* theory of invariants. As he explained to Cayley, [24]

I am bound by usage and by the wants of the undergraduates reading for *mods* to give a course or courses of lectures on *pure* synthetic geometry.

The School of Moderations, or 'Mods' as they were familiarly called, were examinations then taken after four or six terms at Oxford, and they represented a major hurdle toward

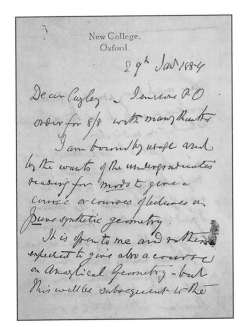

Letter from Sylvester to Cayley, 29 January 1884.

the degree. It was thus important for students to get the 'right' preparation, and Sylvester was clueless as to even what textbook he might draw from.

The realization was quick in coming that he was not going to enjoy the liberties at Oxford that he had at The Johns Hopkins University. No longer would he be able to rove in his lectures where his immediate research interests took him: in fact, his research would ostensibly play no role at all in his Oxford classroom. Moreover, he felt the need to set and keep to a text, while he got himself up to speed with the material. A realist who recognized his own predilections, Sylvester was rightly concerned about whether he would succeed in disciplining himself to this kind of focused teaching.

After opting to sit out his first term in residence, as was his right, Sylvester was engaged in his teaching by May 1884, giving three lectures a week on projective geometry and using a book for his own personal preparation by his Italian friend Luigi Cremona. After a somewhat fitful start, he was able to report to Cayley on 20 May that 'I am getting on now without trouble in my lectures' and 'manage to draw on the board sufficiently well for the purposes of instruction all the geometrical constructions required'.[25] Sylvester's handwriting had always been atrocious and drawing on the board had never been a strength!

The agony (and ecstasy) of an inaugural lecture

His first term over, Sylvester spent much of the summer break away from Oxford agonizing over the inaugural lecture that he was already supposed to have given, but which he had postponed. His anxiety stemmed from the fact that, while he had no idea for a topic, he wanted to make the best possible impression. In May, his new colleague Robert Clifton, Professor of Experimental Physics and Director of the Clarendon Laboratory, had suggested that he speak on Descartes as a geometer, an idea that Sylvester found 'admirable' because 'Descartes' name is very much up just now in Oxford as a Philosopher'.[26] Still, Sylvester was no historian and had no clue how to gather the material necessary for such a lecture. In July, he rued 'the dreadful incubus of a Public Lecture which the University Statutes require of me next term'.[27]

The 1884–85 academic year, Sylvester's first full year in Oxford, found him even more tied down than he had felt thus far. His ongoing indecision about the inaugural lecture and its second postponement got the autumn off to a rocky start, but he was at least initially optimistic about his teaching. He was lecturing three times a week 'to a numerous class on Analytic Geometry', and had taken as his text three chapters out of the German edition and adaptation by Wilhelm Fiedler of the advanced textbook on conic sections by his Irish friend George Salmon.[28] While, as Sylvester confessed to Cayley, 'This is a much more interesting subject to me than Cremona on projections', it was certainly not the sort of material that students would expect to be tested on in Mods.[29] It was invariant-theoretic, and not even a particularly gentle introduction thereto, coming from a textbook in German. Sylvester's choice of subject matter might have made sense for his former graduate classes in Baltimore, but it was not a wise choice for Oxford undergraduates operating within a system of set examinations.

Oblivious, Sylvester was upbeat. As he told Cayley:

Oxford is a dear good mother . . . and stretches out her arms with impartial fondness to take all her children to her bosom even those whom she has not reared at her breast.

In uniquely Sylvesterian style, he even rhapsodized in a poetry recitation to his colleagues at New College in the fall of 1884 about how welcome he felt among them. He intoned:[30]

> Where glows by Wykeham's fane the sacred fire
> And generous hearts the Muses' precincts throng
> A murmur spread that I should wield the lyre
> And strike the chords of my untutored song.

Ten more lines followed in this vein.

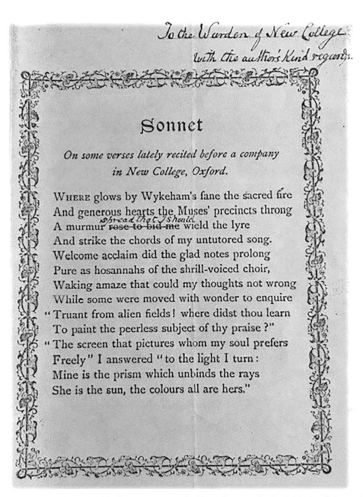

To the Warden of New College
with the author's kind regards.

Sonnet

*On some verses lately recited before a company
in New College, Oxford.*

WHERE glows by Wykeham's fane the sacred fire
And generous hearts the Muses' precincts throng
A murmur ~~rose to bid me~~ *spread that I should* wield the lyre
And strike the chords of my untutored song.
Welcome acclaim did the glad notes prolong
Pure as hosannahs of the shrill-voiced choir,
Waking amaze that could my thoughts not wrong
While some were moved with wonder to enquire
"Truant from alien fields! where didst thou learn
To paint the peerless subject of thy praise?"
"The screen that pictures whom my soul prefers
Freely" I answered "to the light I turn:
Mine is the prism which unbinds the rays
She is the sun, the colours all are hers."

Sonnet 'On some verses lately recited before a company in New College, Oxford'. Notice Sylvester's handwritten emendation in line 3: he was constantly revising his poetry.

By the Hilary Term of 1885, things were not looking so rosy. Sylvester was once again lecturing on analytic geometry from the Salmon–Fiedler text, but at the beginning of the term he was 'in a most wretched mental condition', having begun 'to think that [he] was unequal to [his] post and to weigh whether [he] might not in good faith to resign the professorship'.[31] He was then 70 years old.

Thankfully, the spring brought a brighter mood, with more lecturing on analytic geometry to 'a fairly numerous class' and a supplementary intercollegiate course on matrices at the request of some Oxford tutors.[32] The latter was just the kind of course that Hereford George had argued for before the 1877 Commission, and it was much more similar in spirit to the kind of courses that Sylvester had taught at Hopkins than his Oxford undergraduate lectures in geometry. Nevertheless, unlike his courses in Baltimore, this one did little to stimulate Sylvester's own mathematical creativity. Aside from four short and derivative notes, he accomplished little in the way of research during the 1884–85 academic year, and this, as always, left him despondent.

Things looked more promising in the autumn of 1885. Sylvester had hit on some new mathematical ideas on what he called 'reciprocants' – polynomial expressions in successive derivatives of y with respect to x that remain invariant when x and y are interchanged. These, he had decided, would form the topic of his long-postponed inaugural lecture. On 12 December 1885, when he finally delivered it, the Sylvester of Hopkins days was definitely back, and he was extolling the virtues of his latest work with characteristic hyperbole.

A large audience turned out on that Saturday to hear him speak, with friends like Pieter Schoute, professor of mathematics at the University of Groningen, journeying all the way from the Netherlands, others like Percy MacMahon riding up from London, and still others coming from the various Oxford colleges and elsewhere.[33] Sylvester felt obliged to open with an apology for the tardiness of the lecture:[34]

I have waited, before addressing a public audience, until I felt prompted to do so by the spirit within me craving to find utterance, and by the consciousness of having something of real and more than ordinary weight to impart.

The lecture on Descartes that he had contemplated and agonized over simply had not moved him. It would not have been a lecture of real 'weight', but rather something perfunctory worked up merely to satisfy an obligation. Now, though, he had hit upon something exciting, something 'far bigger and greater, and of infinitely more importance to the progress of mathematical science' than the topics of any of the public lectures he had ever given.[35] While 'no subject during the last thirty years has more occupied the minds of mathematicians, or lent itself to a greater variety of applications, than the great theory of Invariants', he told his audience, the theory of reciprocants 'infinitely transcends in the extent of its subject-matter, and in the range of its applications' ordinary invariant theory.

Sylvester first gave the technical flavour of his subject by means of an example:

$$dy/dx \; \cdot \; d^3y/dx^3 - {}^3/_2\left(d^2y/dx^2\right)^2.$$

As he explained,[36]

if in this expression the x and y be interchanged, its value, barring a factor consisting of a power of the first derivative, remains unaltered, or, to speak more strictly, merely undergoes a change of algebraical sign.

Sylvester called this particular example a 'mixed reciprocant', because it involves the first derivative dy/dx; otherwise, a reciprocant is 'pure'. Geometrically,

every pure reciprocant corresponds to, and indicates, some singularity or characteristic of a curve, and *vice versâ* every such singularity of a general nature and of a descriptive (although not necessarily of a projective) kind, points to a pure reciprocant.

The algebraic isolation of pure reciprocants thus provides geometrical information about the underlying curve, thereby allowing the sort of algebrization of geometry toward which much of Sylvester's mathematical work had been directed.

With his example laid out, Sylvester took the opportunity to reflect publicly on Oxford, his perception of its educational mission, and his view of his role on its faculty. His reflections involved a confession:[37]

During the past period of my professorship here, imperfectly acquainted with the usages and needs of the University, I do not think that my labours have been directed so profitably as they might have been either as regards the prosecution of my own work or the good of my hearers: my attention has been distracted between theories waiting to be ushered into existence and providing for the daily bread of class-teaching.

Not surprisingly, his solution was to abandon his prescribed undergraduate teaching and, as he had done in Baltimore, to offer courses solely on the research in which he was immediately engaged,

and thus, by example, give lessons in the difficult art of mathematical thinking and reasoning – how to follow out familiar suggestions of analogy till they broaden and deepen into a fertilising stream of thought – how to discover errors and to repair them, guided by faith in the existence and unity of that intellectual world which exists within us, and is at least as real as that with which we are environed.

This, he told his listeners, was what he had done in Baltimore, and it had resulted in much new work produced in cooperation with his students. As he characterized their work together:[38]

It was frequently a chase, in which I started the fox, in which we all took a common interest, and in which it was a matter of eager emulation between my hearers and myself to try which could be first in at the death.

At The Johns Hopkins University he had taught his students how to be *researchers*. He wanted to do the same in Oxford.

In short, Sylvester sounded the call, at least in mathematics, to make Oxford a Hopkins, a university with the dual mission of research and the training of future researchers. To this end, he had already announced that his Hilary Term course in 1886 would be on the theory of reciprocants, and declared that, by offering high-level courses and with the aid of a few more men of demonstrated mathematical talent, together with his 'brother Professors and the Tutorial Staff of the University':[39]

we could create such a School of Mathematics as might go some way at least to revive the old scientific renown of Oxford, and to light such a candle in England as, with God's grace, should never be put out.

The sentiments that Sylvester expressed in his inaugural lecture were as self-serving as they were bold. Since his arrival at Oxford, he had been miserable – especially in his

---

Lectures on the Theory of Reciprocants.

By Professor Sylvester, F. R. S., *Savilian Professor of Geometry in the University of Oxford.*

[Reported by J. Hammond.]

LECTURE XI.

We may write for the Annihilator of an Invariant

$$\Omega = a_0\dot{a}_1 + 2a_1\dot{a}_2 + 3a_2\dot{a}_3 + \ldots . ja_{j-1}\dot{a}_j$$

and for its opposite

$$O = ja_1\dot{a}_0 + (j-1)a_2\dot{a}_1 + (j-2)a_3\dot{a}_2 + \ldots . + a_j\dot{a}_{j-1},$$

where the pointed letters $\dot{a}_0, \dot{a}_1, \dot{a}_2, \ldots . \dot{a}_j$ stand for the partial differential operators $\partial_{a_0}, \partial_{a_1}, \partial_{a_2}, \ldots . \partial_{a_j}$. Suppose Ω and O to operate on any function $U(a_0, a_1, a_2, \ldots . a_j)$; then

$$\Omega O U = (\Omega . O + \Omega * O)\, U$$

and

$$O \Omega U = (O . \Omega + O * \Omega)\, U,$$

where the full stop between O and Ω signifies multiplication, and the asterisk operation on the unpointed letters only. Thus,

$$\Omega . O = O . \Omega,$$

and, consequently, $(\Omega O - O \Omega)\, U = (\Omega * O - O * \Omega)\, U.$

Now, $\Omega * OU = \{1 . ja_0\dot{a}_0 + 2(j-1)a_1\dot{a}_1 + 3(j-2)a_2\dot{a}_2 + \ldots . + j . 1a_{j-1}\dot{a}_{j-1}\}\, U,$

and $O * \Omega U = \{1 . ja_1\dot{a}_1 + 2(j-1)a_2\dot{a}_2 + \ldots . + (j-1)\,2a_{j-1}\dot{a}_{j-1} + j . 1a_j\dot{a}_j\}\, U,$

whence we readily obtain

$$(\Omega O - O \Omega)\, U = j(a_0\dot{a}_0 + a_1\dot{a}_1 + a_2\dot{a}_2 + \ldots . + a_j\dot{a}_j)\, U$$
$$- 2(a_1\dot{a}_1 + 2a_2\dot{a}_2 + 3a_3\dot{a}_3 + \ldots . + ja_j\dot{a}_j)\, U.$$

Introducing the conditions of homogeneity and isobarism, viz.

$$(a_0\dot{a}_0 + a_1\dot{a}_1 + a_2\dot{a}_2 + \ldots . + a_j\dot{a}_j)\quad U = iU$$

and

$$(a_1\dot{a}_1 + 2a_2\dot{a}_2 + 3a_3\dot{a}_3 + \ldots . + ja_j\dot{a}_j)\, U = wU,$$

Vol. IX.

One of Sylvester's papers on the theory of reciprocants.

teaching – and yet at the same time he recognized how successful the Hopkins dynamic had been for his mathematical research, as well as for that of his students. What better way to improve his own lot than to take matters into his own hands and publicly redefine his professorship? If the experiment worked, moreover, he would have succeeded in creating a more satisfying professional life for himself, while also making Oxford more of a competitor in the international mathematical arena. It could be a win-win situation.

A Hopkins on the Cherwell?

During that Hilary Term, as Sylvester recounted to Cayley, he had 'a class of 14 or 15' that included 'five or six of the college tutors'.[40] The course continued into the Trinity and Michaelmas Terms of 1886, with some thirty-three lectures in total, ultimately appearing in print over three volumes of the *American Journal of Mathematics*.[41]

In its published version, he fleshed out the skeleton that he had displayed in his inaugural lecture of his theory of reciprocants, and he boasted of the new results that his auditors had produced. Indeed, he noted that papers had 'been contributed on the subject to the *Proceedings of the Mathematical Society of London* by Messrs Hammond, MacMahon, Elliott, Leudesdorf, and Rogers'.[42] James Hammond was a 32-year-old mathematician who had recently moved to Oxford; Percy MacMahon served as a Captain in the Royal Artillery and as an Instructor at the Royal Military Academy; Edwin Bailey Elliott had taken First Class Honours in the Final Mathematical Schools in 1873 and was a tutor at The Queen's College and Lecturer in Mathematics at Corpus Christi College; Charles Leudesdorf had taken his Oxford MA degree in 1876 and was a fellow and lecturer in mathematics at Pembroke College; and Leonard Rogers had taken First Class Honours in Mathematics at Balliol College in 1884 and had remained at Oxford as a tutor until 1888. In the presence of this gifted group of college tutors and mathematics lecturers, Sylvester seemed to be realizing not only his personal goals of stimulating his own research programme and animating a mathematical school at Oxford, but also Hereford George's ideal of research seminars given by the professoriate to nurture the research agendas of the teaching fellows. Perhaps Oxford could be a Hopkins after all.

After three consecutive terms the subject of reciprocants had exhausted itself, although Sylvester had already made the fateful decision to turn to a non-research-inspired topic for his course, 'Surfaces of the Second Order', as illustrated by the new set of models that he had acquired for the University from the German firm of Ludwig Brill.[43] The lectures were not a success. Unlike the course on reciprocants that drew in the tutors

(*Left*) In 1886 Oxford University acquired a set of Ludwig Brill's models of surfaces, used to illustrate geometrical ideas. Sylvester gave a course of lectures on these surfaces and a public lecture on the models.
(*Right*) This announcement, describing a delay in purchasing the models, gives details on the content and organization of Sylvester's Savilian lectures.

because of its potential to stimulate original research, this one, although it was something Sylvester wanted to think about, covered well-known material. Moreover, because the University's regulations required that questions on the mathematical examinations come from more elementary Euclidean geometry, there was absolutely no incentive for undergraduates to attend.

After his teaching triumphs of 1886, this failure devastated him. As he confided miserably to Daniel Gilman:[44]

I am out of heart in regard of my Professorial work in this University in which all the real power of influencing the studies of the place lies in the hands of the College Tutors and in which I can see no prospect of doing any real good. The mathematical school here is at a very low ebb and the number of Mathematical students continually diminishing . . . It depends exclusively on the Tutors whether a Professor can get undergraduates to attend his lectures and their attendance on such lectures may be as irregular as they please. Few of the Tutors recommend such attendance and many not merely discourage but actually prohibit it to those under their control on the ground that it will not pay in the examinations in the schools.

Sylvester had known this, of course. It had been the reason why the attendance in his low-level classes had fallen steadily since his arrival at Oxford, and it was the source of the confession that he had made in his inaugural lecture. He had naively thought that he could declare it best to teach what interested him at the moment, and that the students would come. It had worked with his research seminars of 1886, but it could not work at the undergraduate level, given the power that the tutors had acquired in the wake of the

University's reforms and the well-entrenched system of examinations that was only very slowly changing during Sylvester's Oxford tenure. He felt betrayed, declaring to Gilman:

Entre nous this University except as a school of taste and elegant light literature is a magnificent sham. It seems to me that Mathematical science here is doomed and must eventually fall off like a withered branch from a Tree which derives no nutriment from its roots.

Oxford was never going to be a Hopkins for Sylvester, and now he knew it.

This tough realization proved almost cathartic. The remainder of his tenure in the Savilian chair found him simply getting on with life as he wanted to live it – thinking about mathematics, composing verses, being with friends, travelling, and enjoying the fruits of his accomplishments. He taught the material that he knew he had to teach, but he did develop at least two new courses, on elementary and more advanced number theory, that sparked some new research results on prime numbers.[45]

In what might be seen as one last hurrah, moreover, he initiated an experiment to try to raise mathematical awareness and competence locally. On 29 May 1888 he gathered together Oxford's mathematical lecturers in his rooms in New College and pleaded for the formation of a mathematical society that would provide a venue for the discussion of nascent mathematical ideas – ideas that were perhaps not ready for a meeting of the London Mathematical Society, but that could be honed in the friendlier and more relaxed atmosphere of regular and local get-togethers.[46] With all present in agreement with his plan, Sylvester asked his young colleague and fellow reciprocant-theorist, Edwin Elliott, to send out a circular to all mathematics graduates in Oxford inviting them to attend an organizational meeting on Saturday 9 June. A gratifying three dozen people answered the call to constitute the original membership of what they dubbed the Oxford Mathematical Society.

At this first official meeting, they elected Sylvester as President, Bartholomew Price (the Sedleian Professor of Natural Philosophy and a Fellow of Pembroke College) and William Esson (Fellow of Merton College) as Vice Presidents, and Edwin Elliott as Secretary – and set their rules. Foremost among these was the prohibition on any sort of a society publication. The rationale for this was twofold:[47]

On the one hand slight or tentative papers, which their authors could not think of as enough developed for publications were welcomed, and on the other more ambitious efforts could be talked about without prejudice to their admissibility by, say, the London Mathematical Society.

Edwin Bailey Elliott (1851–1937).

That Sylvester had accurately taken the pulse of Oxford mathematics in 1888 seems clear from the fact, that during the Society's first five years, twenty-three different contributors presented an average of just over three papers at each of the six annual meetings, for a total of just under 100 papers. Of the contributors, Sylvester was unsurprisingly the most prolific with seventeen presentations, but Hammond, Elliott, Leudesdorf, and others also took advantage of the forum to test their new results on an audience of mathematical peers. If Oxford in the 1880s was not yet prepared for the active training of future researchers in its classrooms, it apparently *was* ready for the extracurricular support system that the Oxford Mathematical Society provided for those who nevertheless wanted to contribute to mathematics at the research level. At least in this way, Sylvester *did* help to realize the goal of fostering research and a research ethos in an Oxford that was still undergoing its transformation into a modern university.

Despite this success, the ill health that had been dogging Sylvester finally took its toll. A leg injury in November 1887 had required surgery early in the following year and the appointment of a deputy to execute the teaching duties of the Savilian chair during that Hilary Term. Ongoing eye troubles had required frequent treatment and were ultimately diagnosed as cataracts in both eyes. Depression, from which he had long suffered, recurred repeatedly. In March 1893 a temporary deputy was named, but it took a year before a permanent deputy, the same William Esson who had stood in for Sylvester in

The original members of the Oxford Mathematical Society.

the winter of 1888, was appointed.[48] Michaelmas Term of 1892 was Sylvester's last in an Oxford classroom.

As his melancholy deepened, Sylvester took less and less interest in the correspondence that at times had so thoroughly consumed and defined him. His many friends and supporters learned of his retirement through the press. The Jewish community, which had long held him up as an example in Victorian England, read the news in *The Jewish Chronicle*, while Gilman and his former colleagues at The Johns Hopkins University

James Joseph Sylvester, during his tenure of Oxford's Savilian chair. This portrait by Alfred E. Emslie was commissioned by St John's College, Cambridge. Completed in April 1889, it was exhibited at the Royal Academy in London before being hung in the College Hall.

William Esson (1838–1916), Sylvester's deputy and then successor as Savilian professor.

saw notice of the retirement in one of the other English newspapers. Sylvester's American, English, and other mathematical colleagues heard about it quickly through their well-developed grapevine. After Sylvester died in London in March 1897, William Esson succeeded him as the thirteenth Savilian Professor of Geometry.

Conclusion

Sylvester's appointment to Oxford's Savilian chair had indeed been a dream come true for him, and yet, on returning to England to take up the post, he had entered unfamiliar territory. Oxford was not the Cambridge of his student days, and it was certainly not a Johns Hopkins. It was an ancient university with objectives that differed from those of the new, forward-looking, and modern research university that he had left behind in Baltimore. He had to adjust his sense of 'the professor' and the 'professional mathematician' to the realities of Oxford – its undergraduate thrust and its well-entrenched system of college tutors and set examinations.

During his tenure, however, Sylvester tried in his own idiosyncratic ways to instil a research ethos along the banks of the Cherwell. He had drawn the college tutors and lecturers into his advanced courses on reciprocants. He had spurred them to do original and publishable research of their own on the topic. He had animated the Oxford Mathematical Society. But Oxford in the decade from 1884 to 1894 when Sylvester was its Savilian Professor of Geometry was simply not ready to complete its transition into a modern research university. That would come only in the 20th century.

G. H. Hardy (1877–1947) in his rooms at New College, Oxford.

CHAPTER 6

G. H. Hardy and E. C. Titchmarsh

ROBIN WILSON

This chapter presents the Oxford life and labours of two Savilian professors of the 20th century. The first of these, the successor as Savilian Professor of Geometry to James Joseph Sylvester and William Esson, was G. H. Hardy, who is widely recognized as the most important British pure mathematician of the first half of the 20th century. Although he is usually thought of as a Cambridge man, his years from 1920 to 1931 in Oxford were actually his happiest and most productive. Edward Charles Titchmarsh was Hardy's first research student in Oxford, and later his successor as Savilian Professor of Geometry. He too enjoyed a highly successful career while at Oxford, being recognized for his research achievements and for his widely admired books.

In *Mathematics in Victorian Britain*, the chapter on Victorian Oxford looks to the future:[1]

It was with the arrival of Esson's successor G. H. Hardy as Savilian professor in 1920 that the modern era of mathematics in Oxford really began. In many ways Hardy was a natural heir of Henry Smith in number theory, and he arrived in Oxford at the height of his mathematical powers, picking up essentially where Sylvester had left off by creating a fully fledged research school in mathematics for the first time in Oxford. One needs only to look at the list of subsequent professors, including E. C. Titchmarsh, Sydney Chapman,

Robin Wilson, *G. H. Hardy and E. C. Titchmarsh*. In: *Oxford's Savilian Professors of Geometry*.
Edited by Robin Wilson, Oxford University Press. © Oxford University Press (2022). DOI: 10.1093/oso/9780198869030.003.0006

E. A. Milne, Henry Whitehead, C. A. Coulson, Graham Higman, Sir Michael Atiyah, Sir Roger Penrose, Daniel Quillen, and Simon Donaldson, to see that Oxford has maintained its research momentum in both pure and applied mathematics.

This is a bold claim. Research in mathematics was generally at a low ebb in Oxford when Hardy arrived, and in this chapter we investigate the extent to which a 'fully fledged research school in mathematics' was created during his years as Savilian professor.

G. H. Hardy

As mentioned above, Godfrey Harold Hardy was the most important British pure mathematician of the first half of the 20th century. Although he spent many years at Cambridge University, based at Trinity College as student, lecturer, and fellow, his years from 1920 to 1931 as Savilian Professor of Geometry at Oxford University proved in many ways to be his most successful.

In *A Mathematician's Apology*, Hardy claimed that

I was at my best at a little past forty, when I was a professor at Oxford,

while C. P. Snow, in an introduction to the same book, described Hardy's Oxford years as 'the happiest time of his life'.[2] At Oxford he was at the prime of his creative life, receiving prestigious honours and holding high office in several organizations, and during his years there he wrote over one hundred papers, including many of his most important investigations with his long-term collaborator, J. E. Littlewood.

The chair is advertised

From 1897 to 1916, as we have seen, the Savilian geometry chair had been occupied by William Esson, following the tenure of J. J. Sylvester who had taken up the position at the age of 69 after returning from the United States. Esson was also Sylvester's deputy from 1894 to 1897, when failing eyesight prevented Sylvester from lecturing.

Esson died in 1916, during the First World War, and the consequent election to the vacant Savilian professorship was suspended until the end of hostilities. Eventually, in October 1919, an advertisement appeared in the *Oxford University Gazette*, inviting candidates to apply. The requirements of the post included the following; the annual stipend of £900 is equivalent to about £40,000 today.

The Electors to this Professorship intend to proceed to an election, before the close of the year, of a Professor to come into office on January 19, 1920 . . .

The Professor will be a Fellow of New College, and will receive a stipend of £900 per year, partly from the College and partly from the Savilian endowment and the University Chest.

It will be the duty of the Professor to lecture and give instruction in Pure and Analytical Geometry . . .

He is also bound to give not less than forty-two lectures in the course of the academical year; six at least of such lectures shall be given in each of the University Terms, and in two at least of the University Terms he shall lecture during seven weeks not less than twice a week . . .

The Statutes of the University provide as follows with regard to the Savilian Professors of Geometry and Astronomy : –

Ne alia quapiam professione eodem tempore fungatur Professor alteruter; nec munus Observatoris Radcliviani, nec officium Praelectoris alicujus in quovis Collegio publice legentis cum munere suo conjungat. [Neither professor should fulfil any other profession at the same time. Those who deliver public lectures should not combine with their own duties the duties of an observer at the Radcliffe Observatory nor the office of Praelector in any college.]

Although Hardy was not a geometer, a number of considerations made the position attractive to him. By 1919 he had become disillusioned with his Cambridge home of Trinity College, as administrative duties became increasingly time-consuming and irksome. More notably, he had been sickened by the First World War and by the pro-war attitudes of the other fellows, especially in their treatment of Bertrand Russell, then a young lecturer at Trinity. In 1916 Russell was convicted and fined for writing a pacifist pamphlet, and two years later he was imprisoned for his continued anti-war activities. The College Council voted to remove his lectureship, eventually reinstating it in 1919 once the war was over, but as one of Russell's strongest supporters at Trinity, Hardy found the atmosphere stifling and needed a change of scene.[3]

At the same time, Hardy had been devastated by the departure of his co-worker, the brilliant young Indian mathematician Srinivasa Ramanujan. In early 1913, Ramanujan had written to Hardy from India, enclosing some of his mathematical jottings. Their depth and originality astounded Hardy who invited him to England, and the three-year partnership they enjoyed in Cambridge produced some spectacular papers that caused Littlewood[4] to comment on

Bertrand Russell (1872–1970), Srinivasa Ramanujan (1887–1920), and J. E. Littlewood (1885–1977).

... a singularly happy collaboration of two men, of quite unlike gifts, in which each contributed the best, most characteristic, and most fortunate work that was in him.

Their joint work led to Ramanujan's election to Fellowship of the Royal Society and a Trinity College fellowship. But he became very ill and returned home to India in March 1919 to recover. Working with him in Cambridge had been Hardy's main solace during the bitter war years, and once Ramanujan had left, Trinity seemed empty.

Yet another reason for Hardy's discontent was that his continuing researches with Littlewood were not going well. Their collaboration had begun in 1911 and became probably the most prolific partnership in mathematical history, for they co-authored almost one hundred papers on a wide range of subjects as they dominated the English mathematical scene for the first half of the 20th century. But in late 1919 their collaboration seemed to be in trouble, as Hardy confided to Russell:[5]

I wish you could find some tactful way of stirring up Littlewood to do a little writing. Heaven knows I am conscious of my huge debt to him. But the situation which is gradually stereotyping itself is very trying for me. It is that, in our collaboration, he will contribute ideas and ideas *only*: and that *all* the tedious part of the work has to be done by me. If I don't, it simply isn't done, and nothing would ever get published . . . Really, its got very badly on my nerves. It is the fear that, unless I can get more leisure somehow, I shall work myself to pieces that has driven me, more than anything else, into standing for the vacant Oxford professorship (which I probably shant get, as it is a chair of "Geometry").

Fortunately this difficulty proved to be short-lived, and they produced some of their best joint work in the ensuing years, as we shall see.

In due course this notice appeared in the *Oxford University Gazette*:

SAVILIAN PROFESSORSHIP OF GEOMETRY

At a meeting of the Electors held on Friday, December 12, GODFREY HAROLD HARDY, M. A., Fellow of Trinity College, Cambridge, was elected Savilian Professor of Geometry, to enter on office on January 19, 1920.

The Electors for the Oxford chair were the Vice-Chancellor (the historian H. E. D. Blakiston), the Warden of New College (the Revd. William Spooner, of 'spoonerism' fame), the Waynflete Professor of Pure Mathematics (E. B. Elliott), the Savilian Professor of Astronomy (Herbert H. Turner), and three others, M. J. M. Hill, Percy A. MacMahon, and C. H. Sampson.[6] It is not known who the other candidates were, although Hardy had originally encouraged William Henry Young, the newly appointed professor of pure mathematics in Aberystwyth, to apply, as outlined by Ivor Grattan-Guinness:[7]

Another candidate was W. H. Young, Hardy's senior by fourteen years and equally distinguished in mathematical analysis, who had passed most of his career on the Continent. The two men corresponded on the matter, Hardy offering that "if there were a question between us, my candidature should not be pressed." But soon afterwards he changed his mind, being of course "sorry to have to go back on my word but I must do so" and pleading the apparent Cambridge tendency "to pile up work on the staff to a point which may rapidly become intolerable". Outsider Young, knowing that connections score most points in academic life, withdrew with dignity.

G. H. Hardy was duly appointed and arrived in Oxford early January 1920 in time for the beginning of Hilary Term.

Hardy arrives in Oxford

Hardy's inaugural lecture as Savilian professor took place at the University Observatory on 18 May 1920 and opened with the following words:[8]

I think that a professor should choose, for his inaugural lecture, a subject, if such a subject exists, to which he has made himself some contribution of substance and about which he has something new to say … I have therefore finally decided, after much hesitation, to take a subject which is quite frankly mathematical, and to give a summary account of the results of some research which, whether or no they contain anything of any interest or importance, have at any rate the merit that they represent the best that I can do.

The lecture, in which Hardy referred to contributions of his Savilian predecessors Henry Smith and J. J. Sylvester, was entitled 'Some famous problems of the theory of numbers, and in particular Waring's problem'. This problem, first posed in 1770 by the Cambridge mathematician Edward Waring, attempted to extend Joseph-Louis Lagrange's celebrated 'four-square theorem' that every positive integer can be written as the sum of at most four perfect squares. Claiming that every positive integer can also be written as the sum of at most 9 cubes or 19 fourth powers, Waring asked whether such representations continue for all higher powers, and if so, how many powers are needed.[9]

This occasion marked Hardy's first involvement with Waring's problem, and during his Oxford years he and Littlewood co-authored six papers on the topic in a series entitled 'Partitio Numerorum'; these articles are notable for exploiting the so-called 'Hardy–Littlewood circle method', a powerful technique whose origins lay in a celebrated paper on partitions of numbers that Hardy and Ramanujan had published in 1918.[10]

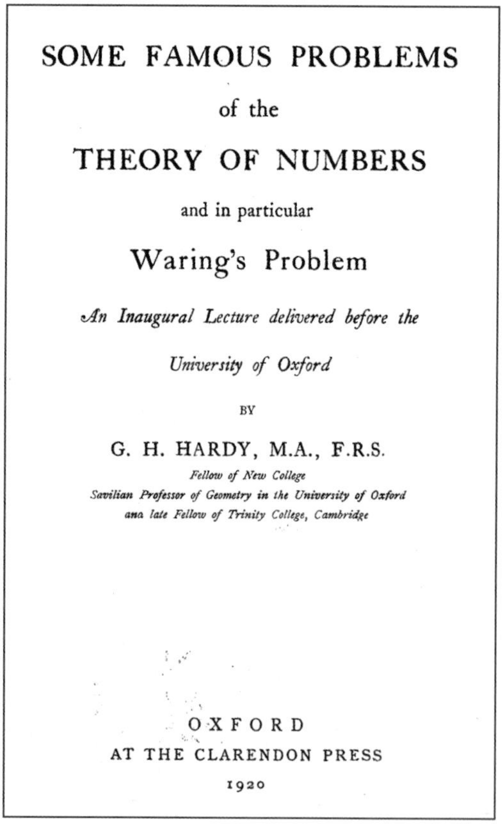

SOME FAMOUS PROBLEMS

of the

THEORY OF NUMBERS

and in particular

Waring's Problem

An Inaugural Lecture delivered before the

University of Oxford

BY

G. H. HARDY, M.A., F.R.S.

Fellow of New College
Savilian Professor of Geometry in the University of Oxford
and late Fellow of Trinity College, Cambridge

OXFORD
AT THE CLARENDON PRESS
1920

G. H. Hardy's inaugural lecture on the theory of numbers.

New College. At New College, the college attached to the Savilian chairs, Hardy felt completely at home. The Wardens during his time there were the aforementioned William Spooner, who lectured on ancient history, philosophy, and divinity, and H. A. L. Fisher, who had served in David Lloyd George's government and later wrote a classic three-volume *History of Europe*. The fellowship included the philosopher Horace W. B. Joseph, the physiologist John S. Haldane, and the conductor and music educator Sir Hugh Allen, while among almost 300 undergraduates were the future politicians Hugh Gaitskell, Richard Crossman, and Frank Pakenham (Lord Longford). The few students who studied mathematics were taught by the applied mathematician E. H. Hayes, who lectured on dynamics; he was succeeded shortly after Hardy's arrival by E. G. C. Poole, an expert on differential equations, who remined at New College for the next twenty years.

Hardy was a true 'College man', particularly enjoying the informality and friendliness of the Senior Common Room. Indeed, Littlewood later claimed that Hardy 'preferred the Oxford atmosphere and said they took him seriously, unlike Cambridge,'[11] while Titchmarsh observed:[12]

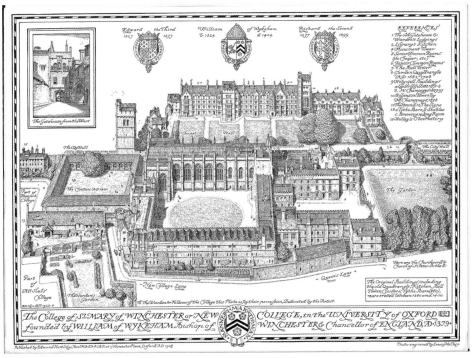

New College, Oxford (engraving by Edmond Hort New, 1923–28); Hardy's rooms were in the Holywell Buildings at the top of the picture.

He was an entertaining talker on a great variety of subjects, and one sometimes noticed everyone in common room waiting to see what he was going to talk about.

and the economist Lionel Robbins likewise remembered:[13]

the leader of talk in the senior common room, arranging with mock seriousness complicated games and intelligence tests, his personality gave a quality to the atmosphere quite unlike anything else in my experience . . . If, as I think, the New College of those days had a very special distinction among institutions I have known, I have little doubt that much of this special quality was evoked by Hardy's catalytic presence.

Mathematical contemporaries. When Hardy arrived at Oxford in early 1920, Mathematics was a sub-faculty of the Physical Sciences with no premises of its own. Here the new 42-year-old Savilian professor joined his aged professorial colleagues, who included the Sedleian Professor of Natural Philosophy (A. E. H. Love), the Waynflete Professor of Pure Mathematics (E. B. Elliott), and the Savilian Professor of Astronomy (H. H. Turner). There were also about a dozen college tutors who taught and lectured in their colleges, including J. E. Campbell (Hertford), T. W. Chaundy (Christ Church), A. L. Dixon (Merton), E. H. Hayes (New College), H. T. Gerrans (Worcester), A. E. Joliffe (Corpus Christi), A. L. Pedder (Magdalen), J. W. Russell (Balliol), C. H. Sampson (Brasenose), and C. H. Thompson (Queen's). Most of these had been talented Oxford undergraduates who progressed to lectureships and fellowships in their own or another college.

Augustus Love, the applied mathematician and geophysicist, occupied the Sedleian chair for over forty years. The Royal Society awarded him its Royal Medal and Sylvester Medal, and he received the prestigious De Morgan Medal from the London Mathematical Society, which he had served as President from 1912 to 1914.[14] His two-volume monumental treatise on the mathematical theory of elasticity was influential, and his Oxford lectures were models of clear thinking and style. He had a widely admired moustache.

Twenty years earlier, Love had taught Hardy at Cambridge, as Hardy recalled:[15]

My eyes first were opened by Professor Love, who taught me and gave me my first serious conception of analysis. But the great debt which I owe to him — he was, after all, primarily an applied mathematician — was his advice to read Jordan's famous *Cours d'Analyse*; and I shall never forget the astonishment with which I read that remarkable work . . . and I learnt for the first time what mathematics really meant. From that time onwards I was . . . a real mathematician, with sound mathematical ambitions and a genuine passion for mathematics.

In pure mathematics the Waynflete chair had been created in 1892, and its first occupant was Edwin Bailey Elliott FRS, a former fellow of The Queen's College who had been President of the London Mathematical Society from 1896 to 1898. An algebraist, his textbooks on invariant theory had been highly regarded, but his mathematics was becoming out of date, he had 'no sympathy with foreign modern symbolic methods', and his teaching was not well regarded. Indeed, the Oxford mathematician and textbook writer W. L. Ferrar was later to write:[16]

Elliott, the great man on Algebra & all its developments, a man who has written books which have put the works of his rivals on the bookshelves, is the worst lecturer who ever picked up chalk. I have spent hours on him last term & not got a scrap of good from him.

Elliott retired as Waynflete Professor in 1921 and was replaced by Arthur Lee Dixon FRS, a former fellow of Merton College, whose work ranged from algebra and geometry to elliptic functions. He made few mathematical innovations, but published several papers on analytic number theory and is remembered as knowing 'some very pretty stuff' in

Two of Hardy's contemporaries from the 1920s: Augustus Love (1863–1940) and A. L. Dixon (1867–1955).

19th-century geometry.[17] A keen sportsman, he could reportedly crack a walnut in the crook of his arm. A popular figure, he became President of the London Mathematical Society from 1924 to 1926, during Hardy's time in Oxford.

Herbert Hall Turner FRS, the Savilian Professor of Astronomy, had been one of the above-mentioned Electors, and like Augustus Love he was a Cambridge man. A leader in the international efforts to produce an astrographic chart of the heavens who latterly worked in seismology, he was a good communicator and on four occasions presented the Royal Institution's Christmas Lectures.

In 1928 two new mathematical chairs, the Rouse Ball professorships, were created in Oxford and Cambridge. Unlike its Cambridge counterpart, where it is reserved for pure mathematics and Littlewood was appointed, the Oxford chair has always been held by applied mathematicians. Its first occupant was Edward Arthur Milne FRS, a former undergraduate student of Hardy's in Cambridge. At Oxford Milne revolution-ized the study of theoretical applied mathematics, and is described in his Royal Society obituary as[18]

one of the greatest figures in the greatest period of development of modern astrophysics and cosmology,

He was later President of the London Mathematical Society from 1937 to 1939 and President of the Royal Astronomical Society from 1943 to 1945. Subsequent holders of Oxford's Rouse Ball chair have included C. A. Coulson and Sir Roger Penrose.

Hardy's research

Before Hardy there had been no flourishing research tradition in Oxford, although Sylvester had tried to initiate one in the 1880s. When Hardy arrived, research activity was generally at a low ebb. Love had produced his best work in the 1890s, Dixon had stopped publishing around 1910, and Elliott had produced little since around 1914. In-deed, as Elliott remarked in a lecture at the Oxford Mathematical and Physical Society's 200th meeting on 16 May 1925:[19]

I still hold soundly that our business as teachers in a University was to educate, to assist young men to make the best use of their powers . . . 'But how about research and original work under this famous system of yours? You do not seem to have promoted it much.' Perhaps not! It had not yet occurred to people that systematic training for it was possible.

But systematic training for research *was* possible. Around this time Oxford's DPhil degree was introduced, and research students could thereafter study for either a BSc degree (roughly equivalent to a modern MSc) or a doctorate. While in Oxford Hardy had seven BSc students, including J. P. Duminy (later a South African test cricketer and Vice-Chancellor of the University of Cape Town), Graham Sutton (who became Director-General of the UK Meteorological Office), and Gertrude Stanley (later head of the mathematics department of Westfield College, London). He also supervised no fewer than sixteen DPhil students, including:

E. C. Titchmarsh (Balliol) who qualified to graduate but never did so,

U. S. Haslam-Jones (Queen's), DPhil 1928,

E. H. Linfoot (Balliol), DPhil 1928,

L. S. Bosanquet (Balliol), DPhil 1929,

M. L. Cartwright (St Hugh's), DPhil 1928,

E. M. Wright (Jesus), DPhil 1931.

Mary Cartwright (later Dame Mary) became Mistress of Girton College, Cambridge, and was the first woman to be President of the London Mathematical Society; Edward Wright (later Sir Edward) collaborated with Hardy on *An Introduction to the Theory of Numbers* and later became Vice-Chancellor of the University of Aberdeen.

On Hardy's success as a research leader and doctoral supervisor, Titchmarsh recalled:[20]

He was always the head of a team of researchers, both colleagues and students, whom he provided with an inexhaustible stock of ideas on which to work. He was an extremely kind-hearted man, who could not bear any of his pupils to fail in their researches. Many Oxford D.Phil. dissertations must have owed much to his supervision.

while Hardy's approach, as remembered by yet another student, was as follows:[21]

If you were a student of Hardy's, he gave you a problem that he was sure you could solve. You solved it. Then he asked you to generalize it in a specific way. You did that. Then he suggested another generalization, and so on. After a certain number of iterations, you were finding (and solving) your own problems. You didn't necessarily learn to be a second Gauss that way, but you could learn to do useful work.

But with many of his Oxford colleagues largely inactive on the research side, Hardy felt the need for a more stimulating environment, as he had enjoyed in Cambridge. So, shortly after his arrival, he introduced regular Friday evening advanced classes in pure mathematics at New College. At these informal gatherings, Hardy would invite interested colleagues and research students, and any mathematical visitors to Oxford, to present

G. H. Hardy and Mary Cartwright. At a time when women were sometimes excluded from lectures, Hardy was always very supportive of women's education.

their current work and discuss the mathematical topics of the day. In his obituary of Hardy, Titchmarsh reminisced about these advanced classes, observing that[22]

Whatever the subject was, he pursued it with an eager single-mindedness which the audience found irresistible. One felt, temporarily at any rate, that nothing else in the world but the proof of these theorems really mattered. There could have been no more inspiring director of the work of others . . .

while many years later, Mary Cartwright recalled:[23]

I felt that I joined a rather very special group when I began to attend Hardy's class at New College in January 1928 . . . I think that we knew that we were studying under a very great man.

Meanwhile, much of Hardy's own research work while at Oxford continued to be with Littlewood.[24] Although they were now resident at different universities, this made no difference to their productivity as they corresponded by letter and succeeded in producing about fifty joint papers in the 1920s. Their collaboration covered a wide range of subjects, including Diophantine approximation, additive and analytic number theory, summability, inequalities, the theory of functions, and Fourier series. A joke of the time, recalled by the Danish mathematician Harald Bohr, held that[25]

Nowadays, there are only three really great English mathematicians: Hardy, Littlewood and Hardy-Littlewood

Hardy and Littlewood at Trinity College, Cambridge. This photograph, taken in the 1920s, is the only surviving photograph showing the two of them together.

while the German number-theorist Edmund Landau added:[26]

The mathematician Hardy-Littlewood was the best in the world, with Littlewood the more original genius and Hardy the better journalist.

Indeed, Hardy himself considered Littlewood to be the more creative partner, with Littlewood generally creating the logical structure of their joint papers and Hardy crafting the final version.[27]

Hardy's teaching

There are three terms in the Oxford academic year: Michaelmas Term (October to December), Hilary Term (January to March), and Trinity Term (April to June). The lecture list for Hilary Term, 1920, presented a range of courses in mathematics, including Elliott on 'A First Course on The Theory of Functions', Turner on 'Elementary Mathematical Astronomy' with some practical work, Love on 'Electricity and Magnetism', Dixon

Subject.	Lecturer.	Time.	Place.
A First Course on the Theory of Functions ...	Waynflete Professor of Pure Mathematics, E. B. Elliott, M.A.	M. W. F. 12 ..	Queen's
Elementary Mathematical Astronomy	Savilian Professor of Astronomy, H. H. Turner, M.A., D.Sc.	M. W. 11	University Observatory.
†Practical Work	,, ,, ,,	To be arranged	,,
Electricity and Magnetism	Sedleian Professor of Natural Philosophy, A. E. H. Love, M.A., D.Sc.	T. Th. S. 10 ..	Electrical Laboratory.
Introduction to the Analytical Geometry of the Plane (last half of Term).	Savilian Professor of Geometry, G. H. Hardy (M.A. Camb.).	To be arranged	To be arranged
Infinitesimal Calculus	J. E. Campbell, M.A.	T. S. 12 ..	Hertford ..
Tridimensional Rigid Dynamics	H. T. Gerrans, M.A.	W. F. 9 ..	Worcester ..
Dynamics of a Particle and Rigid Dynamics ..	C. H. Thompson, M.A.	M. W. F. 11	Queen's ..
Hydrostatics	A. L. Dixon, M.A.	Th. 11 ..	Merton ..
Elementary Analysis	A. E. Jolliffe, M.A.	M. F. 12 ..	Corpus Christi
Curve Tracing	J. W. Russell, M.A.	T. Th. 10 ..	Balliol ..
Theory of Equations	,, ,, ...	S. 9.30–11 ..	,, ..
Elementary Theory of Plane Curves	C. H. Sampson, M.A.	W. F. 10 ..	Brasenose ..
Problems in Pure Mathematics	T. W. Chaundy, M.A.	T. Th. 12 ..	Christ Church
Elements of Dynamics of a Particle and of a Rigid Body.	E. H. Hayes, M.A.	M. F. 11 ..	New College ..
Subjects of Preliminary Examination (Mathematics) and Group C. (1).	A. L. Pedder, M.A.	T. Th. F. 5 ..	Magdalen ..

I.—MATHEMATICS.

Part of the mathematics lecture list for Hilary Term, 1920.

(perhaps surprisingly) on 'Hydrostatics', and other topics covered by the college tutors, usually within their own colleges as there was then no central Mathematical Institute.[28] Although Hardy had arrived in Oxford in early 1920, it was not until halfway through Hilary Term that he began a series of lectures entitled 'Introduction to the Analytical Geometry of the Plane'; these took place on Tuesdays and Saturdays at 11 a.m. at New College.

Hardy was not a geometer, but he assiduously lectured on the subject each term, as required by the statutes of his professorship. In Trinity Term he concluded the course he had begun in Hilary Term, and in the following Michaelmas Term he presented 'Analytic Geometry' and 'Applications of Analysis to Geometry'. In later years he lectured on 'Solid Geometry' and 'Elements of Non-Euclidean Geometry', as well as branching out into non-geometric topics, such as 'Chapters in the Theory of Functions', 'Theory of Numbers', 'Calculus of Variations', and 'Famous Mathematical Problems'. Hardy's Advanced Class in Pure Mathematics' began to appear on the lecture lists from Michaelmas Term 2020, and in 1922 he also presented a course on 'Elements of Mathematics for Philosophers', which attracted a large attendance.

In *A Mathematician's Apology*, Hardy confessed that[29]

I hate 'teaching', and have had to do very little, such teaching as I have done having been almost entirely supervision of research; I love lecturing, and have lectured a great deal to

SAVILIAN PROFESSOR OF GEOMETRY:
G. H. HARDY, M.A.
The Professor will lecture this Term as follows :—
(1) Analytical Geometry, on Tuesdays and Saturdays
at Noon (but not on Thursdays as previously
announced).
(2) Applications of Analysis to Geometry, on Thursdays at Noon, beginning on October 21. In
connexion with this lecture an Advanced Class
in Mathematics will be held on Fridays at
8.45 P.M.
All the lectures and classes will be given at New
College.

An announcement in the *Oxford University Gazette* of Hardy's geometry lectures for Michaelmas Term, 1920.

extremely able classes; and I have always had plenty of leisure for the researches which have been the one great permanent happiness of my life.

Indeed, Hardy was widely admired as a brilliant lecturer. The physicist Freeman Dyson, who attended some of Hardy's later lectures in Cambridge, enthused:[30]

He lectured like Wanda Landowska playing Bach, precise and totally lucid, but displaying his passionate pleasure to all who could see beneath the surface ... And each lecture was carefully prepared, like a work of art, with the intellectual dénouement appearing as if spontaneously in the last five minutes of the hour. For me these lectures were an intoxicating joy ...

But Hardy could be a controversial figure, especially in his views on teaching and examining, as noted by the historian of science Jack Morrell in his book on *Science at Oxford 1914–1939*:[31]

For Hardy an examination could do little harm provided its standard was low. He also attacked the colleges who, greedy for firsts, encouraged tutors to stunt mathematical education by absorption in examination technique, and to tire themselves out in trying to turn a comfortable second into a marginal first. He thought most Oxford tutors did twice as much teaching as any active mathematician should be asked to undertake.

Several of Hardy's lecture courses have been preserved in notes by the Oxford undergraduate E. H. Linfoot, later one of his doctoral students. These two pages are from his 1924–25 course on the theory of numbers.

Other preoccupations

Throughout the 1920s Hardy had many preoccupations besides his research and teaching, being heavily involved with international issues and with several national societies. But first he suffered a personal loss.

The death of Ramanujan

In the years prior to his coming to Oxford Hardy had produced some spectacular joint work with Ramanujan. But Ramanujan had become ill and returned to Madras in 1919. He died on 26 April 1920, at the age of 32, shortly after Hardy's arrival in Oxford, and Hardy was devastated:[32]

For my part, it is difficult for me to say what I owe to Ramanujan – his originality has been a constant source of suggestion to me ever since I knew him, and his death is one of the worst blows I have ever had.

In 1921 Hardy wrote a beautifully crafted obituary notice which appeared in the *Proceedings of the London Mathematical Society* and the *Proceedings of the Royal Society*, and was described by the *Manchester Guardian* as 'among the most remarkable in the literature about mathematics'.[33] For the next three years he wrote several papers based on, and extending, Ramanujan's work, while carefully working through his friend's extensive notebooks and papers, and in 1927, with Ramanujan's long-time Indian friend P. V. Seshu Aiyar and the English mathematician B. M. Wilson, he published the *Collected Papers of Srinivasa Ramanujan*.[34] Here, commenting on Ramanujan's 'profound and invincible originality', Hardy observed:[35]

He would probably have been a greater mathematician if he had been caught and tamed a little in his youth; he would have discovered more that was new, and that, no doubt, of greater importance. On the other hand he would have been less of a Ramanujan, and more of a European professor, and the loss might have been greater than the gain.

Hardy's internationalism

In the years following World War I, Hardy felt cut off from Continental mathematics and from his many friends and colleagues overseas. Even before his arrival in Oxford he had been cooperating with peacemaking efforts by the Swedish mathematician Gösta Mittag-Leffler, and had written to him in January 1919 insisting that[36]

all scientific relationships should go back precisely to where they were before. I have never valued any recognition more than what I have obtained in Germany; and I should regard the loss of my personal relationships with German mathematicians as an irretrievable calamity.

In order to re-establish connections he travelled widely, to Scandinavia and Germany in particular, lecturing and visiting mathematical colleagues wherever he went. He also corresponded with his German number-theorist friend Edmund Landau about his views on the war; Landau replied, agreeing with Hardy's opinions but 'with trivial changes of sign'.[37]

In another letter to Mittag-Leffler,[38] Hardy discussed the quadrennial International Congresses of Mathematics, insisting that he was

in no circumstances prepared to take part in, subscribe to, or assist in any manner directly or indirectly, any Congress from which, for good reasons or for bad, mathematicians of particular countries are excluded.

In particular, he boycotted the International Congresses that took place in Strasbourg in 1920 and in Toronto in 1924, because German, Austrian, and Hungarian mathematicians had not been invited. He attended the 1928 Congress in Bologna, after such restrictions had been lifted.

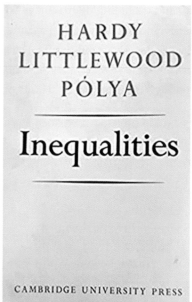

(*Left*) G. H. Hardy and George Pólya in Oxford in the 1920s.
(*Right*) Hardy, Littlewood, and Pólya's *Inequalities*.

Throughout the 1920s Hardy enjoyed welcoming foreign visitors to Oxford as a way of restoring international connections and encouraging research activity in the University. These included L. E. Dickson, the American algebraist and historian of number theory, who stayed with him in the autumn of 2020 after attending the Strasbourg Congress, and Harald Bohr from Denmark.

From 1923 to 1928 the International Education Board of the Rockefeller Foundation initiated a scheme of one-year fellowships that enabled distinguished scholars in the natural sciences to visit institutions in other countries.[39] Hardy strongly supported this project and became very involved with organizing its academic programme.

The first mathematician to become an international Rockefeller Fellow was the Hungarian George Pólya who was then based in Zürich. On Hardy's recommendation Pólya visited England in 1924–25, spending half of the year with him in Oxford and the other half with Littlewood in Cambridge; an outcome was a classic book on *Inequalities*, authored by the three of them.[40]

Two other international scholars who spent time in Oxford under the Rockefeller scheme were the German logician Wilhelm Ackermann and the Russian algebraist and number-theorist Alexander Ostrowski. Yet another was Abram Besicovitch from Russia, who visited Hardy in Oxford and for whom Hardy secured a teaching post at the

G. H. Hardy (*left*) leads his cricket team at the Oxford British Association meeting in 1926: from left to right are H. K. Salvesen, W. L. Ferrar, W. J. Langford, E. C. Titchmarsh, E. H. Neville, E. H. Linfoot, and L. S. Bosanquet.

University of Liverpool from 1926 to 1927. Besicovitch then transferred to Cambridge University, where he lectured for many years and in 1950 succeeded Littlewood as the Rouse Ball Professor of Mathematics.

Honours and societies

While in Oxford, Hardy received many prestigious honours and appointments in recognition of his outstanding contributions to mathematics.

Royal Society. Hardy had been elected a Fellow of the Royal Society of London in 1910, at the early age of 33. Ten years later, shortly after his arrival in Oxford, he received the Society's Royal Medal, awarded annually for 'the most important contributions to the advancement of natural knowledge'.

British Association. As we saw in Chapter 4, the British Association for the Advancement of Science began life in 1831 with a meeting in York and another in Oxford in the following year. In 1922 its annual meeting took place in Hull, and Hardy was appointed President of the Mathematics and Physics Section (Section A), giving his Presidential address on 'The Theory of Numbers'.[41] Four years later, when the meeting was again held in Oxford, Hardy indulged his favourite pastime of playing cricket by leading a team that he dubbed 'Mathematicians v. The Rest of the World'.[42]

National Union of Scientific Workers. Hardy became president of this Union from 1924 to 1926 and made recruiting speeches for them. His left-wing leanings emerge clearly from the following remarks to an audience of scientists:[43]

although our jobs are very different from a coalminer's, we are much closer to coalminers than capitalists. At least we and the miners are both skilled workers, not exploiters of other people's work, and if there's going to be a line-up, I am with the miners.

Mathematical Association. The Mathematical Association has always been concerned with the teaching and learning of mathematics. Founded in 1871 as the Society for the Improvement of Geometrical Teaching, and with J. J. Sylvester as its President during the year 1891–92, it received its present name in 1894. H. H. Turner was its President from 1909 to 1911, and Hardy occupied this position from 1924 to 1926.

Each year Hardy was required to give a presidential address, and the first of these, in 1925, was entitled 'What is geometry?', in which he insisted that[44]

I do not claim to know any geometry, but I do claim to understand quite clearly what geometry is.

He then went on to discuss this question at some length from mathematical, philosophical, and educational points of view.

Hardy had always detested the over-demanding and over-technical Cambridge examinations, known as the Tripos, and in particular its listing of all the candidates in

---

224. [M¹. 8. g.] *A curious imaginary curve.*

The curve
$$(x+iy)^2 = \lambda(x-iy)$$
is (i) a parabola, (ii) a rectangular hyperbola, and (iii) an equiangular spiral. The first two statements are evidently true. The polar equation is
$$r = \lambda e^{-3i\theta},$$
the equation of an equiangular spiral. The intrinsic equation is easily found to be $\rho = 3is$.

It is instructive (i) to show that the equation of any curve which is both a parabola and a rectangular hyperbola can be put in the form given above, or in the form
$$(x+iy)^2 = x \text{ (or } y\text{)},$$
and (ii) to determine the intrinsic equation directly from one of the latter forms of the Cartesian equation.   G. H. HARDY.

Hardy once joked that the only result that he had ever proved in geometry concerned 'A curious imaginary curve', in which he showed that if a rectangular hyperbola is a parabola, then it is also an equiangular spiral.

numerical order; he had himself taken the Tripos exams a year early and was thereafter labelled as 'Fourth Wrangler'. In his second presidential address for the Mathematical Association, in 1926, Hardy argued 'The case against the Mathematical Tripos', strongly recommending its abolition.[45]

In this connection, an amusing story relates that, as part of his campaign against the Cambridge Tripos, Hardy arranged for George Pólya to sit the Tripos papers during his year in England as Rockefeller Fellow, expecting them to be so unrelated to modern mathematics that Pólya would surely perform poorly. To Hardy's great chagrin, Pólya's excellent performance on these papers would have placed him at the top of the list.[46]

London Mathematical Society. Of all the societies with which Hardy was associated, the one for which he felt the greatest loyalty and affection was the London Mathematical Society. From 1917 to 1926 he was on its Council as Secretary, and became President from 1926 to 1928, during which he gave his presidential address on 'Prolegomena to a chapter on inequalities' where he proudly recalled:[47]

This Society has always meant much more to me than any other scientific society to which I have belonged. My record of attendances, since I became secretary in 1917, has no blemish; I have been at every meeting both of the Council and of the Society, and have sat through every word of every paper.

In 1929 Hardy was awarded the Society's highest prize, the De Morgan Medal. Ten years later he was reappointed as President, the only person to hold this position on more than one occasion.

Hardy visits America

As part of his travelling, Hardy visited the United States for the academic year 1928–29, spending most of the time at Princeton University on an exchange, funded partly by the Rockefeller Foundation, with the American geometer Oswald Veblen who came to Oxford and lectured on 'Tensors and other differential invariants'. In a letter from Princeton to a New College colleague, Hardy confided:[48]

This country is in many ways the only one in the world – it has its deficiencies (tea γ, paper β–, undergraduates rarely above 10 in succulence, dinner rather an uninteresting meal, and at 6.30). But American football knocks all other spectacles absolutely flat: the sun shines more or less continuously: and for quietness and the opportunity to be your own master I've never come across anything like it . . . Everybody seems just at first to be very unsophisticated, but that is a delusion (just as Oxford youth is very much less sophisticated than it seems to be). I have at last learnt what the Oxford manner is – there are 3 or 4 Englishmen here and here it sticks out all over.

While in America, Hardy visited various institutions, including Harvard University, the University of Chicago, The Ohio State University, and Lehigh University, and spent most of February and March of 1928 at the California Institute of Technology (Caltech).

Publications

In addition to more than three hundred papers, book reviews, and obituaries,[49] Hardy wrote a number of books. These included four Cambridge Tracts on *The Integration of Functions of a Single Variable*, *Orders of Infinity*, *The General Theory of Dirichlet's Series* (with Marcel Riesz), and *Fourier Series* (with W. W. Rogosinski).[50] Much better known

The 1967 edition of Hardy's *A Mathematician's Apology*.

G. H. Hardy at New College in 1930.

are *A Course of Pure Mathematics*,[51] a highly influential work that was first published in 1908 and ran to ten editions, and *An Introduction to the Theory of Numbers*,[52] written in 1938 with his former Oxford doctoral student, Edward Wright.

Hardy's best-known publication was *A Mathematician's Apology*, a skilful attempt to explain to a general readership what mathematics (and pure mathematics, in particular) is all about. Written in 1940 at the beginning of World War II, and following a heart attack in the previous year, it is a rather sad retrospection by a mathematician coming to terms with the realization that his mental and physical powers were declining.[53] A revised edition, with a lengthy introduction by the well-known writer C. P. Snow, was published in 1967.

In 1926, to complement its *Proceedings* which had been published since the Society's beginning in 1865, Hardy successfully persuaded the London Mathematical Society to found a new publication, the *Journal of the London Mathematical Society*, with funding from the Rockefeller Foundation. Four years later he re-launched the *Quarterly Journal of Mathematics* in Oxford, following the death of the Cambridge mathematician James Glaisher who had been its editor for over fifty years. Hardy encouraged his colleagues and

INDEX TO VOL. I

The first issue of the *Quarterly Journal of Mathematics*, 1930.

friends to contribute papers, and the first issue featured a bevy of mathematical 'stars', including Littlewood, Cartwright, Dixon, Milne, Pólya, Titchmarsh, and Veblen.

An Oxford School of Mathematics?

At the beginning of this chapter we queried the extent to which a 'fully fledged research school in mathematics' was created during Hardy's years as Savilian professor. There had certainly been many signs of improvement during the 1920s: the joint work of Hardy and Littlewood was widely admired; there were regular gatherings for those interested in research; research students were beginning to earn doctorates and take up academic posts; there was the stimulus of international visitors, and new people (such as E. A. Milne) were arriving; and several Oxford mathematicians played important roles in the London Mathematical Society and were Fellows of the Royal Society.

But Hardy felt that there was still a long way to go. In an important and regularly cited article in the *Oxford Magazine* of 5 June 1930,[54] he championed the cause of Oxford mathematics, bemoaning the fact that the subject 'probably attracts less notice in Oxford than in any comparable university' and that mathematics and physics are 'overshadowed in Oxford not merely by the literary schools but even by other sciences'.

After admitting that the Oxford School had been overshadowed by its Cambridge counterpart, he went on to assert that 'Oxford has produced a good many more distinguished mathematicians than is generally realised', and that 'English opinion has overrated very grossly the status of Cambridge mathematics'. He then declared:

I do not think that it ought to be very difficult for the Oxford School of Mathematics to eradicate this rather humiliating inferiority complex . . . there is no subject in which it would be easier to develop a fine school than in mathematics, if once we had decided that we really wanted to do so . . .

and argued further that

If Oxford wishes to have a first-rate School of Mathematics, it can have one; the difficulty for us is to persuade the University (or rather the Colleges) that such a school is an asset worth possessing. It is a question, primarily, of buying the men; mathematicians are reasonably cheap, but they cannot be had for nothing . . .

Noting that every Cambridge college had at least one mathematician, and that some had three or four, he complained that 'Exeter, Lincoln, Magdalen, Oriel, Pembroke, Trinity, Wadham, and Worcester have not a mathematician among them'. He concluded:

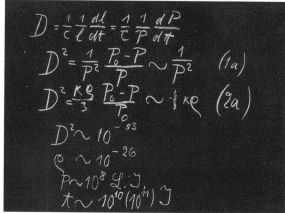

In May 1931, just before Hardy's departure, Albert Einstein visited Oxford to receive an honorary D.Sc. degree and deliver three lectures on relativity; a blackboard from the second one ('The Cosmological Problem') has been preserved. Merton College arranged a special dinner for Einstein, which Hardy, Love, Dixon, and Milne attended.

We are, therefore, asking for a Mathematical Institute as a centre for Oxford mathematical life. It is an obviously reasonable, and a modest, demand, remarkable only because it has been delayed so long, and sooner or later it must be met. Whether it is met sooner or later, the proposal is in itself a most encouraging sign, as evidence of a new spirit of enterprise and assertiveness on the part of Oxford mathematicians, and proof that they are beginning to rebel against the subjection, unworthy of a splendid subject, in which they have acquiesced too long.

It would be some time before a fully fledged Mathematical Institute was in place.[55]

Hardy leaves Oxford

Although he was happy at Oxford, Hardy chose to resign his Oxford chair and return to Cambridge, following the death in 1931 of E. W. Hobson, Cambridge's Sadleirian Professor of Mathematics. Cambridge University remained the mathematical hub of England, and in spite of his efforts to establish a research school in Oxford, Hardy aspired to hold this prestigious Cambridge position. Moreover, he was now aged 54, and if he remained at New College he would eventually have to yield up his rooms upon retirement, whereas he could live at Trinity College for life – as indeed did Littlewood. In a letter to Oswald Veblen, Hardy agonized:[56]

I was much preoccupied with the problem of whether I should send in my name as a candidate for Hobson's chair at Cambridge. I really wanted your advice, for the sake of your well known sanity of judgement, very badly: it was an uncommonly difficult decision to make. In the end I decided Yes; with the result that I leave Oxford this October. The crucial reason was – the opportunity (oddly enough, at least) – of really letting myself go a little in lectures. Also, of escaping from a number of very third rate research pupils (2 of the 5 I have here this year: one really good, 2 so-so, 2 dreadfully trying).

Hardy was successful in securing the Sadleirian chair, and a notice duly appeared in the *Oxford University Gazette* inviting applications for his successor. By this time the stipend of the Savilian professor had increased to £1200 per year, and the Statutes now recognized the importance of research by requiring that

The duties of every Professor shall include original work by the Professor himself and the general supervision of research and advanced work in his subject and department.

Hardy continued to return to Oxford for several weeks every summer to enjoy the atmosphere, revisit his mathematical colleagues, and captain the New College Senior Common Room cricket team in their annual match against the college servants and the choir school. He also returned in Hilary Term, 1936, to give the first lecture to the Invariant Society, Oxford University's undergraduate mathematics society, which still flourishes; this was on the subject of 'Round numbers' (those with few different prime factors).[57]

Hardy's notes on 'Round numbers', the topic on which he gave the first Invariant Society lecture in Hilary Term, 1936.

Hardy retired from Cambridge's Sadleirian chair in 1942. As mentioned earlier, he became President of the London Mathematical Society for the second time, from 1939 to 1941, and served again as its secretary from 1941 to 1945. During his career he had received honorary degrees from Harvard, Athens, Sofia, Oslo, Marburg, Edinburgh, Manchester, and Birmingham, and in 1947 he was elected a foreign member of the Paris Académie des Sciences, one of only ten people (selected from all countries and scientific disciplines) to receive this honour at any one time.

But by this time Hardy's health was rapidly declining, and when he was informed that the Royal Society proposed to present him with its highest accolade, the Copley Medal, he commented:[58]

Now I know that I must be pretty near the end. When people hurry up to give you honorific things there is only one conclusion to be drawn.

He died on 1 December 1947, the very day that he was due to receive the Copley Medal. As his epitaph, one can hardly improve on these few words from *A Mathematician's Apology*:[59]

I still say to myself when I am depressed, and find myself forced to listen to pompous and tiresome people, 'Well, I have done one thing *you* could never have done, and that is to have collaborated with both Littlewood and Ramanujan on something like equal terms.'

Edward Charles Titchmarsh

On 19 March 1964 a memorial meeting was held at the Mathematical Institute in Oxford to celebrate the life and labours of Edward (Ted) Titchmarsh, who had died unexpectedly in the previous year. Mary Cartwright opened the proceedings by speaking on functions of a complex variable, and was followed by other distinguished figures who described Titchmarsh's work on Riemann's zeta-function in number theory, Fourier (and other) transforms, and eigenvalue expansions. These were all topics to which he had made major contributions.

In 1931 Titchmarsh had become Hardy's successor as Savilian professor, but it all happened by chance.[60] In the summer of that year he was visiting Oxford to examine a doctoral candidate. While there, he bumped into W. L. Ferrar, the mathematics tutor at Hertford College, who asked him whether he had applied for Hardy's vacant position. Titchmarsh said no, but (encouraged by Ferrar) thought that he might do so. At the last moment he submitted a brief application on a single sheet of paper, saying that he wished to apply for the Savilian chair but that he could not undertake to lecture in geometry. Two days later he received a telegram from Hardy saying that he had been elected, and the specification for the post was subsequently relaxed so that its occupant was no longer required by statute to lecture on 'pure and analytic Geometry', but instead on 'Geometry or some other branch of Pure Mathematics'.

Writing to Oswald Veblen about the appointment, Hardy commented:[61]

Titchmarsh got my job (I was not an elector, but Littlewood was). I fancy L. would have preferred Besicovitch; but I expected the electors, with the opportunity of taking a genuinely first-rate O. [Oxford] product, to do so. The man I am unhappy about is Mordell: first cut out from Camb. by my decision to stand there, and then here by the O. candidate.

School and college years

Titchmarsh's father was a Congregational Minister – first in Newbury in Berkshire, where Ted was born, and later in Sheffield, where from the age of 9 the boy attended King Edward VII School. As he later recalled:[62]

At school I normally got good marks in most subjects. The first occasion on which I distinguished myself in mathematics was when I was, I think, in one of the fourth forms. The headmaster for some unknown reason made the whole upper school do an arithmetic paper, the same for all forms. The result was that the mathematical specialists in the sixth form came out top, and I came next . . . it had become clear that mathematics was my real subject, and I began to specialise in it.

While at school he also developed a passion for playing and watching cricket, like his Uncle Charles who played for Hertfordshire and the Marylebone Cricket Club.

In December 1916 Titchmarsh won an Open Mathematical Scholarship to Balliol College, going up to Oxford in October 1917 for just one term. He then left for almost two years of war service, as Second Lieutenant in the Royal Engineers (Signals), and from August 1918 was posted to France as a dispatch rider, on horse and then motorcycle.

Titchmarsh returned to Oxford in October 1919, being tutored at Balliol by J. W. Russell. As Mary Cartwright wrote in her obituary of him:[63]

at Russell's first lecture the room was packed to the doors, and Russell said: "Ah, there's my clever pupil Mr Titchmarsh, he knows it all, he can go away." Russell's methods were extremely regular; he dictated his lectures word by word, and at tutorials selected portions

Edward C. Titchmarsh (1899–1963).

of book work and examples were handed out, and then, if necessary, solutions to examples. Some of Titchmarsh's solutions replaced the official ones.

A fellow student, Edward Copson, later Professor of Mathematics at St Andrews University, remembered that[64]

He and I took Mods. & Finals at the same time, and he was one of my earliest friends at Oxford. We met in J. W. Russell's tutorial room in Balliol, he a young officer back from the war and I a very juvenile schoolboy . . . when I was in difficulties, I would walk round to Balliol in sure confidence that Titchmarsh would help me. And he always did; in a most kindly and friendly way, he resolved my stupid misunderstandings of my mathematics. I shall always remember him for his kindness and friendliness.

Titchmarsh's undergraduate career was indeed very successful, leading to a First Class Honours Degree and to both the Junior and Senior Mathematical Scholarships.

The 1920s

The 1920s were important years for Titchmarsh, as Hardy became his most important influence in pure mathematics:[65]

I was however principally influenced by G. H. Hardy. From him I learnt what mathematical analysis is, and at his suggestion I devoted myself to research in pure mathematics.

In particular, Hardy's weekly evening mathematical gatherings were of great interest to Titchmarsh, Mary Cartwright, and others, as were the meetings of the Oxford Mathematical and Physical Society which reached its 200th meeting in 1925. The Oxford school in analysis that Hardy had begun during these years would later be developed by Titchmarsh when he succeeded to the Savilian professorship.

In 1923, after a year's research with Hardy in which he also acted as his part-time secretary for correspondence, Titchmarsh was appointed to a senior lectureship at University College, London; there he gave undergraduate lectures, supervised postgraduate students, and began to publish research papers in major journals. In the same year he became a Prize Fellow by Examination at Magdalen College, Oxford, a position that was awarded to graduates of outstanding merit; he held this position for seven years, and was thereby able to keep in touch with Oxford mathematics by regular visits.

Meanwhile, his father had become a church minister in Essex and Edward fell in love with Kathleen, the church secretary's daughter. Kathleen called him 'Oxford's most B.M.'

Titchmarsh's letter of appointment to the Savilian chair.

(brilliant mathematician), while he described her as 'the best girl who ever took up with a poor wandering mathematician and tried to make him into a human being'.[66] They were married in 1925 and had three daughters.

In 1929 Titchmarsh was appointed Professor of Pure Mathematics at the University of Liverpool. Here he became President of the Liverpool Mathematical Society (affiliated to the Mathematical Association) and in 1931 was elected to a Fellowship of the Royal Society. While in Liverpool he also wrote his first book, a Cambridge tract on the Riemann zeta-function, which he expanded twenty years later into a classic text, *The Theory of the Riemann Zeta-function*.[67]

A letter from Hardy to Titchmarsh.

On his election to the Savilian chair in 1931, Titchmarsh returned to Oxford and became a fellow of New College. Here he played a full role, serving on the Audit and Finance Committee for many years, and at one stage taking over the position of Sub-Warden which he carried out quietly and efficiently. A man of few words, it was said of him:[68]

He was a retiring person and not easy to know, but his shyness was relieved by a sense of humour which would suddenly transform him

and that

His silences were benevolent and never oppressive. To his colleagues he was best known for his short but effective interventions in the deliberations of the Governing Body . . .

He also succeeded Hardy as captain of the Fellows' Cricket XI in its annual match against the choir school.

In the world of Oxford mathematics, Titchmarsh's contemporaries included Augustus Love, Sydney Chapman, and George Temple in the Sedleian chair, A. L. Dixon,

Edward Titchmarsh, with Charles Coulson and George Temple.

Henry Whitehead, and Graham Higman as Waynflete professors, and E. A. Milne and C. A. Coulson in the Rouse Ball chair.

As a senior mathematician in the department, Titchmarsh was expected to play an important role in its research activities. A colleague, Edward Thompson, later recalled:[69]

During the early years of his tenure of the Savilian Chair, nearly all postgraduate research in pure mathematics in Oxford was supervised by him, and throughout his life he continued to attract a flourishing group of research pupils. He was a notably successful supervisor, with a nice judgment of a pupil's capability and great skill in selecting a problem which would extend but not defeat.

Shortly after Titchmarsh had returned to Oxford, the Department moved into rooms in the Radcliffe Science Library, and then in 1953 into 10 Parks Road. From then on he was the Curator of the new Mathematical Institute. With a rather old-fashioned view of life he had no telephone in his room, abhorred external noise, and rarely used departmental headed writing paper or dictated a letter, and his manuscripts were invariably handwritten. But he ran the department quietly and efficiently, explaining his role in the following terms:[70]

You may like to know about the sort of set up which we have for mathematics at Oxford. There are four professors – two pure and two applied, and two readers (i.e. minor professors) and a number of university lecturers, all housed in a rather grim Victorian building in Parks Road. This we call the Mathematical Institute. There we carry out what is called 'advanced study and research' as distinct from the main teaching of undergraduates, which is done by the college tutors in their own colleges. This building hums with higher mathematics, and is full of advanced students writing their dissertations for the degree of Doctor of Philosophy . . . It is not exactly a department, for we are all more or less independent. But I am the Senior Professor and Curator of the Institute. It is an honorary post. We all pursue our own lines of thought though these sometimes intersect. This system, or lack of system, works very happily.

Publications, honours, and awards

During his thirty years in Oxford, Titchmarsh's research publications appeared at a great rate, gradually shifting their focus from number theory and trigonometric series to Fourier transforms, eigenfunction expansions, analysis for physicists, and the theoretical background to relativistic quantum mechanics.

It was also at this time that he wrote 'a series of books of quite outstanding merit'.[71] His first major work, *The Theory of Functions* on real and complex analysis, was published in

1932, shortly after his arrival in Oxford.[72] Based on his lectures in London and Liverpool, it was a highly successful and greatly admired work that was later described as[73]

the rebellion of a young, widely read professor against the narrow range to which mathematical analysis was then so often confined.

Titchmarsh's method of writing his books was unusual. After researching on a topic for a number of years, he would 'sign off' with a book that represented a synthesis of his own discoveries. Later writings included a classic text on the theory of Fourier integrals (1937) in which he gave credit to Hardy:[74]

I worked on the theory of Fourier integrals under his guidance for a good many years before I discovered that this theory has applications in applied mathematics, if the solution of certain differential equations can be called applied

and in his two-part book on *Eigenfunction Expansions Associated with Second-order Differential Equations* (1946/1958), he explained his attitude, as a pure mathematician, to a subject that was normally regarded as 'applied':[75]

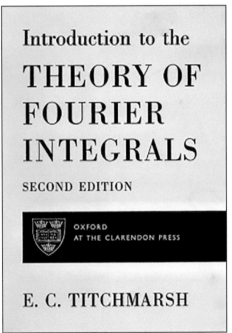

Two of E. C. Titchmarsh's books.

The whole work is the result of an attempt by an "analyst" to understand those parts of quantum mechanics which can be regarded as exercises in analysis. The subject is, however, pursued without much regard to the interests of theoretical physicists. It seems that physicists do not object to rigorous proofs provided that they are rather short and simple. I have much sympathy with this point of view. Unfortunately it has not always been possible to provide proofs of this kind.

Although by nature shy and introverted, Titchmarsh was an enthusiastic advocate for his subject. In 1948 he published a popular *Mathematics for the General Reader*,[76] in which he explored questions that might be posed to mathematicians, such as:

What good is it to know that every number is the sum of four squares? Why do you want to know about prime-pairs? What does it matter whether pi is rational or irrational?

In reply, he explained:

Pure mathematicians do mathematics because it gives them aesthetic satisfaction which they can share with other mathematicians. They do it because for them it is fun, in the same way perhaps that people climb mountains for fun. It may be an extremely arduous and even fatal pursuit, but it is fun nevertheless. Mathematicians enjoy themselves because

Edward C. Titchmarsh in later years.

they do sometimes get to the top of their mountains, and anyhow trying to get up does seem to be worth while.

In 1959 Titchmarsh recorded for the BBC Third programme a twenty-minute radio talk on the nature of mathematical proof, which he then wrote up as an article for *The Listener*.[77]

Through his many publications, Titchmarsh's achievements became widely recognized. He was President of the London Mathematical Society from 1945 to 1947, and received its De Morgan Medal in 1953. The University of Sheffield awarded him an honorary doctorate in 1953, and in the following year he was an invited plenary speaker at the International Congress of Mathematicians, held in Amsterdam. In 1955 the Royal Society awarded him its Sylvester Medal.

Conclusion

Then, in January 1963, suddenly and prematurely, he died in his armchair at home. There were many tributes, as everyone was very fond of him. His professorial colleague Charles Coulson wrote:[78]

There were many things about Ted that I have always much admired – his utter humility, which never betrayed anything but the greatest simplicity; his complete integrity . . . and his kindness to me when I arrived first; to his students (who worshipped him) and to everyone.

Another colleague, Bryce McLeod, recalled:

I shall never forget the quick warm smile with which he would greet me at any time that I went to see him. It was never accompanied by much in the way of words, but I was made to feel nonetheless that he was there and ready to help me in any way that he could. Above all, the quality which I associate with him is that of scrupulous integrity. He would not tolerate anything that was slip-shod anywhere else in his life either. At the same time, there was always apparent in him a very real sympathy for those whose lesser abilities prevented them from reaching his own high standards.

But perhaps we should leave the final tribute to a young visitor to his house who, on being told later that he had been playing dominoes with a great mathematician, remarked:[79]

Well, he didn't seem like it.

Michael Atiyah at his desk in the Mathematical Institute, 1974.

CHAPTER 7

From Michael Atiyah to the 21st century

FRANCES KIRWAN

This chapter looks at the four holders of the Savilian chair of geometry in the second half of the 20th century – Michael Atiyah, Ioan James, Richard Taylor, and Nigel Hitchin – concentrating mainly on the first of these. Michael Atiyah was Savilian professor for only six years, from 1963 to 1969, but he spent most of the following two decades in Oxford as a Royal Society professor, and his impact on the field of mathematics, in Oxford, in the UK, and internationally, was immense.

Michael Atiyah

According to Mary Cartwright, E. C. Titchmarsh was a man of few words but 'benevolent silences'.[1] His successor, Michael Atiyah, recounted that when he arrived in Oxford in 1961 to take up a Readership, he had to visit Titchmarsh as Curator of the Mathematical Institute to get his office key:[2]

I was duly ushered into his big room where he was sitting at his desk. I sat down and he handed over the key, and I then expected a speech of welcome or some words of advice, but we just sat in silence. After five minutes I left.

Frances Kirwan, *From Michael Atiyah to the 21st century*. In: *Oxford's Savilian Professors of Geometry*. Edited by Robin Wilson, Oxford University Press. © Oxford University Press (2022). DOI: 10.1093/oso/9780198869030.003.0007

Michael Atiyah was an entirely different character, who would not have sat in silence for even five seconds on meeting another mathematician for the first time. Graeme Segal, one of Atiyah's first research students in the 1960s, recalled that he 'overflowed with energy and enthusiasm' and 'poured forth a torrent of ideas and advice'.[3] Exactly the same description fits the Michael Atiyah that I met when I became his student nearly two decades later.

The research interests of the 20th-century Savilian professors of geometry were not necessarily focused on what modern mathematicians would regard as 'geometry'. Indeed, as we saw in Chapter 6, Titchmarsh agreed to take up the position only on condition that the requirement to give lectures each year on geometry was relaxed. Michael Atiyah, on the other hand, was one of the greatest geometers of the century. He believed very strongly in the unity of mathematics, and always felt it very important to keep pure and applied mathematics together, but he was sure that he was a geometer. He once commented:[4]

Algebra is the offer made by the devil to the mathematician. The devil says: 'I will give you this powerful machine, it will answer any question you like. All you need to do is give me your soul: give up geometry and you will have this marvellous machine'

– and earlier, and less provocatively,[5]

Broadly speaking I want to suggest that geometry is that part of mathematics in which visual thought is dominant whereas algebra is that part in which sequential thought is dominant. This dichotomy is perhaps better conveyed by the words "insight" versus "rigour" and both play an essential role in real mathematical problems . . .

We should aim to cultivate and develop both modes of thought . . . geometry is not so much a branch of mathematics as a way of thinking that permeates all branches.

Michael Atiyah held the position of Savilian professor for only six years (1963–69), before spending three years at the Institute for Advanced Study at Princeton and then returning to Oxford as a Royal Society Research Professor until 1990. However, over the three decades from 1961 to 1990 he was arguably the most influential figure in mathematics in the UK, and was one of the leading mathematicians worldwide.

Early years

Michael Francis Atiyah was born in London in 1929. His maternal grandparents were Scottish, but his father Edward Atiyah was Lebanese. Michael's paternal grandfather had trained at the American University of Beirut, although he spent much of his working life in Sudan as a doctor and then a civil servant in Khartoum. Like his father, Edward Atiyah worked for the government of Sudan for two decades; he then settled in England after the end of World War II, where he wrote books, broadcast for the BBC's Arabic service and became secretary to the London office of the Arab League. According to his son (who resembled him closely in this and other ways), he was reputed to have the loudest voice in London at the time.

Michael described Edward's dream to go to Oxford (he graduated with a degree in history from Brasenose College) as follows:[6]

He wanted to convert himself into an Englishman. It didn't quite work out. When he came back to Sudan he found he wasn't part of the English class structure, he was regarded as one of the lower classes, although he was Oxford-educated and regarded himself as culturally English. That turned him over a bit. He became an Arab nationalist to some extent. All his life was divided between wanting passionately to be English and yet sympathising with the Arab political position within the British Empire.

At Oxford, Edward Atiyah had met Robert Levens, the brother of Michael's mother Jean, and became friendly with the Levens family. After their marriage in 1928, Jean and Edward spent their honeymoon in Florence, and they named their first child after Michelangelo. Michael had a sister, Selma, and two brothers, Patrick and Joseph; Patrick became Professor of English Law at Oxford, and Joseph worked in information technology after a mathematics degree at Cambridge.

Michael attended school in Khartoum, where the family was based, for much of his childhood, although during the four months of summer they would usually travel to visit relatives in the UK and the Middle East. The start of his secondary schooling was disrupted by World War II; he spent time in Lebanon and back in Khartoum before being awarded a scholarship to Victoria College in Cairo, where he travelled by train and Nile boat to join classes two years ahead of his age.

By this time Michael's mathematical abilities were already becoming clear. In 1945, having completed his Higher School Certificate exams at the early age of 16, he went to Manchester to live in a lodging house and attend Manchester Grammar School in preparation for the Cambridge University entrance examination. After being awarded a major scholarship to Trinity College, he spent two years undertaking his National Service as a clerk in the regimental headquarters of the Royal Electrical and Mechanical Engineers, and then began his undergraduate career in Cambridge in 1949.

A student in Cambridge

The mathematicians who supervised Michael's undergraduate studies at Trinity were both highly respected Fellows of the Royal Society: the analyst Abram Besicovitch (who would play eight games of chess simultaneously with his students) and the algebraic geometer John A. Todd, whose textbook on *Projective and Analytical Geometry* Michael had studied with enthusiasm during his National Service. It seems that Todd resembled Titchmarsh in some ways, as Michael's biographical memoir of him for the Royal Society recalls:[7]

I remember supervisions with him in the early 1950s quite vividly. He was efficient and helpful in dealing with technical mathematical difficulties; in fact he prided himself on being able to solve all questions in the Mathematical Tripos. But, when these had been dealt with, an awful silence would descend and could last unbroken until the end of the supervision period. A few experiences of this sort meant that I always came with additional questions or topics of conversation that could fill the dreaded silence. My experience was typical: all my friends had to adopt the same defensive tactics.

During one of the lectures attended by Michael Atiyah as a second-year undergraduate, Todd remarked on the lack of a geometrical proof for an algebraic result. Michael

John A. Todd (1908–94) and William Hodge (1903–75).

was able to provide such a proof, and this led to his first publication. He was very appreciative of Todd's encouragement and help, and was much attracted to his style of algebraic geometry. Nonetheless, he decided to choose as his doctoral supervisor another algebraic geometer, William Hodge, who was deeply involved in bringing together new techniques coming from topology, differential geometry, and algebra. In an interview in 1984, Michael recalled:[8]

I'd come up to Cambridge at a time when the emphasis in geometry was on classical projective algebraic geometry of the old-fashioned type, which I thoroughly enjoyed. I would have gone on working in that area except that Hodge represented a more modern point of view – differential geometry in relation to topology; I recognized that. It was a very important decision for me.

At the end of his first year as a graduate student, Michael wrote a prize-winning essay on ruled surfaces which was the basis for another paper; this helped his confidence and enthusiasm for mathematics, as did attending in 1954 the quadrennial International Congress of Mathematicians in Amsterdam. Thirty years later, he observed:[9]

I've been to every International Congress since 1954, I think; the benefit I have derived from those is very mixed.

The first one, which I attended as a young student, was great. I had a chance to hear Hermann Weyl give a talk and it was a tremendous psychological boost. I felt I was one of a large community of several thousand mathematicians. I didn't understand most of the talks. I'd go to them and be lost. I don't think I gained anything in terms of concrete mathematical understanding, but the psychological boost was substantial.

Now as I get older . . . I go out of a sense of duty – I have functions to perform – to talk to people, to give lectures . . . to help people from countries outside the small circuit of very active mathematical countries.

Lily Atiyah

Another joint paper with William Hodge led to Michael's election to a Research Fellowship at Trinity College in 1954 and made possible his marriage in the following year to Lily Brown. Lily was born in 1928, the daughter of a dock worker at the Rosyth naval shipyard in Edinburgh. For a celebration of her life and work at Edinburgh University in July 2018, Michael wrote:[10]

From a humble working-class background she had, by ability and force of character, climbed the educational ladder ... She collected prizes and scholarships to pay her way, defeated the boys in the traditionally male subject of mathematics and graduated in 1949 with a First Class Honours degree.

From Edinburgh University, Lily went to Girton College in Cambridge:

Distinguishing herself in Parts II and III of the Mathematical Tripos, Lily was then taken on by Mary Cartwright as a PhD student ... one of the first cohorts of women graduate students at Cambridge University.

At Cambridge Lily found herself in a group of 'Scottish exiles', including James Mackay (much later Lord Chancellor, but at that point a mathematician at Trinity with Michael Atiyah), who was to become their best man. Lily later recalled how she first met Michael through the Cambridge students' mathematical society:[11]

At a meeting of the Archimedeans I came across a little man sitting on a table with his legs not even reaching the ground.

Later, Michael became secretary of the Archimedeans while Lily was president.

Lily was a year ahead of Michael, and so by the time they married in 1955 she already had a doctorate, a published paper, and a job as an assistant lecturer at Bedford College in London. However, Michael had been awarded a Commonwealth Fellowship to visit the Institute for Advanced Study in Princeton. After Lily's death, at the end of their 62-year marriage, Michael expressed regret that in those days she had been expected to give up her university post on acquiring a husband who travelled the world.

Lily's doctoral supervisor, Mary Cartwright, avoided the potential obstacle to her career of acquiring a husband. She became the Mistress of Girton College in 1948, a year before Lily's arrival there, and would later be appointed a Reader in the Theory of Functions in the University. As we saw in Chapter 6, Mary Cartwright had studied at Oxford and was one of G. H. Hardy's former students, although Titchmarsh supervised her for a year when Hardy was away in Princeton. She was the first female mathematician to be elected Fellow of the Royal Society, as well as the first woman to serve on its Council and to receive its Sylvester Medal, and was also the first female President of the London Mathematical Society.

After their year in Princeton, Michael and Lily returned to Cambridge, where Michael became an Assistant Lecturer and a Tutorial Fellow at Pembroke College in 1958 while Lily did some teaching for Girton. In 1961 they moved to Oxford where Lily taught for

Lily Atiyah and Mary Cartwright.

St Hugh's College; Michael was first a Reader at the Mathematical Institute, and then succeeded Titchmarsh as Savilian Professor of Geometry in 1963. He resigned his Savilian chair in 1969 to become a professor at the Institute for Advanced Study, so he and Lily moved back to Princeton. Three years later, Michael returned to Oxford University as a Royal Society Research Professor. Lily combined bringing up their three sons, John, David, and Robin, with a long and successful career as a teacher at Headington School, a girls' school not far from the Atiyahs' Oxford home. They returned to Cambridge in 1990 on his appointment as Master of Trinity College.

When Michael retired in 1997 they moved to Lily's home city of Edinburgh, where he became an honorary professor at Edinburgh University. Here they lived in a flat within easy reach of the Scottish holiday home that they had built many years earlier in Perthshire, inspired by a stay in a log cabin while driving across Norway to the Stockholm International Congress in 1962. Michael loved gardening and was proud of the trees he had planted in Perthshire. He also loved hill-walking in Scotland and the opportunity it provided for quiet thought; in a cutting I still have of an article from the *Edinburgh Herald* when Michael's 80th birthday was celebrated, he told the interviewer:[12]

While you go for a long walk in the hills or you work in your garden, the ideas can still carry on. My wife complains, because when I walk she knows I am thinking about mathematics.

Michael's career never involved much undergraduate teaching, and in particular, his position as Royal Society Research Professor had no teaching duties. However, he was a wonderful supervisor of doctoral students, of whom he had many, and Lily was very supportive of us all. Graeme Segal recalled that 'he always fired us up, and made us sure we could succeed'.[13] According to Jill Strang, wife of American mathematician Gil Strang and mother of three boys of similar ages to Michael and Lily's sons, the arrival of Michael Atiyah also fired up children's birthday parties![14]

The great collaborator

The award of a Commonwealth Fellowship allowed Michael Atiyah to visit the Institute for Advanced Study in Princeton for the academic year 1955–56. Princeton at the time was an exciting melting pot of new ideas, and this was a crucial year for his mathematical development. As well as meeting many inspiring mathematicians, including the then-recent Fields Medallists Kunihiko Kodaira and Jean-Pierre Serre, he began lifelong friendships with his three main future collaborators: Raoul Bott, Isadore Singer ('Is'), and Friedrich Hirzebruch ('Fritz'). These collaborators came from around the world: Hirzebruch was German and Singer was American, while Bott was of Austrian–Hungarian descent; growing up in Czechoslovakia, and ending up stateless at the end of World War II (as did Michael's own father), Bott was given citizenship by Canada and eventually became a US citizen.

Michael's collaboration with Hirzebruch led to the invention of 'K-theory', which was his first outstanding achievement. Collaboration with Singer led to the 'Atiyah–Singer

Michael Atiyah and Fritz Hirzebruch.

Two of Atiyah's collaborators: Isadore Singer (1924–2021) and Raoul Bott (1923–2005).

index theorem', their most celebrated result for which they were jointly awarded the Abel Prize in 2004. Various later incarnations of the index theorem in the 1970s involved other collaborators – in particular, Bott – and further work with him resulted in a flurry of important papers on gauge theories, moment maps, and equivariant cohomology in the late 1970s and into the 1980s. In a speech that he prepared for the conferment on him in 1984 of the Antonio Feltrinelli Prize by the Accademia Nazionale dei Lincei in Italy, Michael wrote:[15]

most of my work has been carried out in close and extended collaboration with mathematical colleagues. I find this the most congenial and stimulating way of carrying on research. The hard abstruseness of mathematics is enlivened and mollified by human contact.

In 'A personal history' in 2004, he recalled:[16]

My collaborations with Fritz, Raoul and Is led to 9, 13, 16 joint papers respectively, beginning in 1959 and lasting until 1984. I think it is fair to say that for the 25 most productive years of my life these collaborations were the dominant feature. In each case our backgrounds were somewhat different and I benefited accordingly, picking up expertise in the best way possible – by working with an expert!

After Michael's year in Princeton, Hirzebruch was appointed to a chair in Bonn and in 1957 started the 'Mathematische Arbeitstagung', which continues to this day, even after its founder's death. In its early years the Arbeitstagung was an annual informal gathering in Bonn of mathematicians from across Europe and North America, who met to share with each other the most recent research ideas and advances, and to begin or continue collaborations. Many young mathematicians who were to become famous in the second half of the 20th century participated regularly. The programme of talks was decided at the beginning of the meeting, rather than in advance, although the tradition developed that Michael Atiyah would always give the opening lecture. In 1962 K-theory was the new development that dominated the meeting, and in addition to the research talks there was plenty of time for informal discussions, whether at the meeting or during a boat trip along the Rhine.

When asked about communicating mathematics, in an interview in 1984, Michael replied:[17]

ideally as you are trying to communicate mathematics, you ought to be trying to communicate understanding. It is relatively easy to do this in conversation. When I collaborate with people, we exchange ideas at this level of understanding – we understand topics and we cling to our intuition.

It was this communication of mathematical ideas that happened so successfully at the Arbeitstagung every year.

Three decades in Oxford

In 1960 the Waynflete chair at Oxford University became vacant after the unexpected death of the topologist Henry Whitehead, and Michael Atiyah applied for the position. He was unsuccessful: the algebraist Graham Higman, who was already a Reader in Oxford, became the new Waynflete professor. Michael then applied successfully for the Readership previously held by Higman, and moved to Oxford in 1961. Two years later, when Titchmarsh died and K-theory had begun to make a big impact, the Savilian chair was offered to Michael.

Although Michael moved to Oxford only after Whitehead's death, he had already interacted with Whitehead and wrote about his influence on mathematics (as well as his other interests, which included farming pigs).[18] Michael's claim that people's personalities have a major effect on their influence, and that Henry Whitehead had a remarkable and exuberant personality, could equally well be applied to Michael himself. According to Michael, the topology community was very important to Whitehead, who said to him on one occasion: 'Wouldn't it be terrible if one day I woke up and had a brilliant idea in functional analysis?'[19] – his worry being that he would then have to drop all his topology friends and meet an entirely new group of people. This was not a problem

Henry Whitehead (1904–60), topologist and pig farmer.

for Michael: in the two decades after his arrival in Oxford, when his proof of the index theorem took him in a new direction, from algebraic geometry and topology towards analysis, he 'just made a bigger crowd of friends'.[20]

Michael was famous in the mathematical community for giving inspiring seminars and lectures which drew together ideas from disparate sources to make a hugely appealing and convincing whole. He believed that[21]

the informal style of a lecture has much to commend it. Too often mathematics in print is heavy, formal and pedantic, making impossible demands on the reader.

It was not always easy, however, to recollect all the details of his arguments, and undergraduates in both Cambridge and Oxford found it difficult to make head or tail of his lectures. When I began lecturing in Oxford, I was told of a convention that all first-year mathematics undergraduate lectures were never to be given by experts in the relevant subject area; this apparently dated from the time that Michael gave an introductory course to newly arrived undergraduates, and by the end of term was imparting advanced third-year material to a completely mystified audience. As a Royal Society Research Professor from 1972, he was no longer expected to give undergraduate lectures, and this was probably a blessing for all concerned.

He did not always have greater success with talks that were aimed at a more general mathematical audience. Robin Wilson remembers attending such a lecture when Michael stunned everyone with his opening sentence:[22]

I'm going to start with a really down-to-earth example: take two spheres in 7 dimensions and knot them together.

Michael Atiyah always enjoyed lecturing, whether at a conference or in the classroom.

Nonetheless, Michael felt strongly (at least, in his later life) that universities are institutions that are both educational and involved in research, and that they must try to balance the two activities, insisting that[23]

When universities give courses for educational purposes, they should be sure that they are performing the right task for the students, not just giving courses in (say) advanced topology because they are interested in turning out research students. That's a disastrous mistake.

Geometry and physics

From the mid-1970s, Michael Atiyah's interests began to move towards the interactions between geometry and physics. The mathematical physicist Roger Penrose, who had been a research student with Michael in Cambridge, became Oxford's Rouse Ball professor in 1973, and Michael's discussions with him and his student Richard Ward provided one stimulus. Another was his discovery with Isadore Singer and Nigel Hitchin that the index theorem could be used to solve a problem for which many physicists needed to know the answer – the number of instanton parameters.

A third stimulus occurred in 1977 when he met the young theoretical physicist Edward Witten at MIT:[24]

I immediately realised that he was picking up what I was saying much more quickly than the older generation. I was impressed and subsequently invited him to Oxford.

When Witten was awarded a Fields Medal at the 1990 International Congress in Kyoto, Michael wrote his citation:[25]

Graeme Segal (*left*) and Edward Witten.

Although he is definitely a physicist (as his list of publications clearly shows) his command of mathematics is rivalled by few mathematicians, and his ability to interpret physical ideas in mathematical form is quite unique. Time and again he has surprised the mathematical community by a brilliant application of physical insight leading to new and deep mathematical theorems.

Interaction with Witten on such ideas led to the development, by Michael and his former student and collaborator Graeme Segal, of the concept of a topological quantum field theory, which was hugely influential over the succeeding three decades. In a memorial tribute, Witten summarized Michael's importance in redefining the relationship between mathematics and physics:[26]

Atiyah, along with colleagues such as Bott and Singer, played an enormous role in introducing new ideas and encouraging and teaching physicists to study quantum field theory from new points of view . . . His vision and clairvoyance have had a truly far-reaching influence.

In his acceptance speech for the Feltrinelli Prize in 1984, Michael had expressed similar sentiments:[27]

By training, experience and inclination I am a pure mathematician: I find mathematics beautiful, fascinating and irresistible. However, I recognize that its importance arises from its applications in the natural and social sciences and that it depends on the real world for the stimulus of problems and ideas. For these reasons I am delighted that my work has recently brought me in touch with physicists and that it has, in a small way, helped to strengthen the links between mathematicians and physicists.

Later years

As has already been remarked, Michael Atiyah left Oxford in 1990 for Cambridge to become (simultaneously) the Master of his alma mater, Trinity College, the first Director of Cambridge's new Isaac Newton Institute for Mathematical Sciences, and the President of the Royal Society. He always had many interests outside mathematics – in particular, he thought that academics and scientists should become involved in political life, and to some extent he began to increase his own involvement. A committed pacifist and anti-nuclear campaigner, he became president of the Pugwash Conferences on Science and World Affairs for five years at the end of the 1990s. He was also happy to reflect on the nature of mathematics and research in mathematics in interviews and general lectures, and many of these can be found in the seven volumes of his published works. In particular, he considered that[28]

mathematics can I think be viewed as the *science* of *analogy* and the widespread applicability of mathematics in the natural sciences, which has intrigued all mathematicians of a philosophical bent, arises from the fundamental role which comparisons play in the mental process we refer to as 'understanding'.

He believed above all in the unity of mathematics, and to a secondary extent in the importance of simplicity:[29]

The most useful piece of advice I would give to a mathematics student is always to suspect an impressive sounding theorem if it does not have a special case which is *both* simple *and* non-trivial.

In a long interview with *The Mathematical Intelligencer* in 1984, he observed:[30]

I work a lot with people and I think that that's my style. There are various reasons, one of which is that I dabble in a number of different areas. My interest is in the fact that things in different subjects interact; it's very helpful to work with other people who know a bit more about something else and complement your interest. I find it very stimulating to exchange ideas with other people . . .

Some people may sit back and say, "I want to solve this problem" . . . "How do I solve this problem." I don't. I just move around in the mathematical waters, thinking about things, being curious, interested, talking to people, stirring up ideas; things emerge and I follow them up . . . I have never started off with a particular goal, except the goal of understanding mathematics.

He continued:

It is very hard to separate your personality from what you think about mathematics. I believe that it is very important that mathematics should be thought of as a unity. And the way I work reflects that; which comes first is difficult to say. I find the interactions between the different parts of mathematics interesting. The richness of the subject comes from this complexity, not from the pure strand and isolated specialization.

In this interview Michael Atiyah went on to consider such philosophical and social issues as why we do mathematics, and if the reason is because we enjoy doing so, why we should then get paid for it. His answer centred around the view that mathematics is a part of the general scientific culture, and that even if the piece of mathematics with which one is currently involved has no direct relevance or usefulness to others, it contributes to a whole collection of ideas. Moreover, if we believe that mathematics is an integrated subject, with each part potentially contributing to every other, then we are all contributing to a common goal: even if we do not contribute directly to applied mathematics, our work can still be useful to those who do. As to why we should be paid for doing it, he continued:

Everybody has to try to justify his life philosophically, to himself at least. If you are teaching you can say "Well, my job is to teach, I turn out educated young people and I am paid for that. Research I do in my spare time and they allow me to do that out of generosity." But if you're a full-time researcher, then you've got to think much harder about justifying your work.
    In some sense I still do mathematics because I enjoy doing it. I'm glad that people pay me to do what I enjoy. But I try to feel that there is a serious side to it which provides a justification.

Later in the same interview, when asked whether he could know if a mathematical result was true even if he had no proof, Michael answered in the following way, describing a viewpoint that had served him well until then:

If I'm interested in some topic then I just try to understand it; I just go on thinking about it and trying to dig down deeper and deeper. If I understand it, then I know what is right and what is not right . . .
    Broadly speaking, once you really feel that you understand something and you have enough experience with that type of question through lots of examples and through connections with other things, you get a feeling for what is going on and what ought to be right. And then the question is: How do you actually prove it? That may take a long time.

The Index Theorem, for example, was formulated and we knew it should be true. But it took us a couple of years to get a proof. That was partly because different techniques were involved, and I had to learn some new things to get the proof, in that case several proofs.

He recalled a theorem that he had proved, but at first couldn't really understand *why* it was true. For the proof to work, a great many steps along the way all had to be correct; if one link of the chain were to break, then the whole edifice would collapse. After worrying for several years, fearing that if he did not completely understand it, then the result might not even be true, he eventually discovered a completely different type of proof that made the whole issue completely clear.

He concluded by contrasting his approaches to lecturing and writing:

If I give talks, I try always to convey the essential ingredients of a topic. When it comes to writing papers or books, however, then it is much more difficult . . . In papers I try to do as much as I can in writing an account and an introduction which gives the ideas. But you are committed to writing a proof in a paper, so you have to do that.

Michael Atiyah received numerous honours in his long life. He was knighted in 1983 and received the Order of Merit in 1992. He was awarded a Fields Medal in 1966 and the Abel Prize (jointly with Singer) in 2004. A Fellow of the Royal Society from the early age of 33, he served as its President from 1990 to 1995. He was a Foreign Member of the US National Academy of Science and a long list of corresponding bodies in other countries, ranging from France, Germany, and Russia, via China and India, to Australia and Venezuela. He also had an equally long list of honorary degrees from worldwide universities: these included the American University of Beirut, where his grandfather had studied and with which he developed strong links in his later years.

(*Left*) At one time there were three Fields Medallists in Oxford: from left to right: Michael Atiyah, Simon Donaldson, and Daniel Quillen.
(*Right*) Isadore Singer and Michael Atiyah receive the 2004 Abel Prize from King Harald of Norway.

Ioan James and his bust in the Mathematical Institute at Oxford.

Ioan James

Michael Atiyah's successor as Savilian Professor of Geometry was a year older than Michael. Born in 1928, Ioan Mackenzie James attended St Paul's School in London, before completing an undergraduate degree in mathematics as a student at The Queen's College, Oxford, followed by a doctoral degree in algebraic topology under the supervision of the Waynflete Professor of Pure Mathematics, Henry Whitehead.

Like his predecessor in the Savilian chair, Ioan James was awarded a Commonwealth Fellowship, which allowed him to spend the year 1954–55 in the United States, visiting Princeton University and the University of California, Berkeley. He then held a Research Fellowship in Cambridge before returning to Oxford in 1957 as Reader in Pure Mathematics. When Michael Atiyah resigned from his Savilian professorship in 1969 to become a professor in Princeton, Ioan James was appointed as his successor. He remained in this post until his retirement in 1995.

Ioan James was a topologist, following in the footsteps of Henry Whitehead, with whom he collaborated on several papers in homotopy theory and whose collected works he edited. However, just as Michael Atiyah called himself a geometer tending slightly to the direction of topology, Ioan might have been called a topologist tending slightly towards geometry.

Michael Atiyah later recalled how the motivation for his development of K-theory was provided by Ioan James, who told him about a problem that would demonstrate the use of such a new theory:[31]

It was very much an accident in some ways. I was interested in what Grothendieck had been doing in algebraic geometry. Having gone to Bonn I was interested in learning some topology. I was interested in some of the questions that Ioan James had been studying on topological problems related to projective spaces. I found that by using Grothendieck's formulas these things could be explained, one got nice results. There was Bott's work on the periodicity theorem; I knew him and his work.

He later added:[32]

In fact, there was a third important component that was necessary to produce the motivation, namely some problem which would demonstrate that such a new theory would be of use. This came from Ioan James (my colleague in Cambridge at the time) who was working on stunted projective spaces. Playing around with the formulae I soon discovered that one would get strong results in homotopy theory on James's problem. This made it seem worth while to develop the appropriate machinery to make this formal, and K-theory grew out of that.

Before his unexpected death in 1960, Henry Whitehead had approached Robert Maxwell, the chairman of Pergamon Press, with a proposal to found a new journal called *Topology*. By the time its first issue appeared, Whitehead had died, but Ioan became an editor in 1962 and continued in this role for many decades. Ioan was also heavily involved with the London Mathematical Society, as its Treasurer for ten years from his appointment to the Savilian chair, and then as its President from 1984 to 1986. He was awarded the Society's Junior Berwick Prize in 1959 and the Senior Whitehead Prize in 1978, and was elected a Fellow of the Royal Society in 1968.

Ioan James has written a number of books on topology, but after he retired from the Savilian chair in 1995, his interests moved towards the history of mathematics and science. He edited and contributed to the *History of Topology* in 1999, and later published a biographical memoir of the Princeton topologist, James Alexander. In 2002 he published a collection of sixty short biographies of *Remarkable Mathematicians: From Euler to von Neumann*, and went on to publish companion volumes on physicists, biologists, and

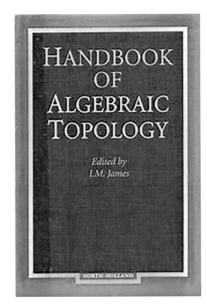
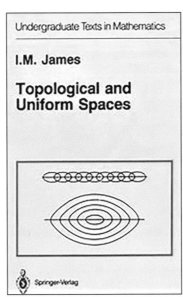

Two of Ioan James's topology books.

engineers.[33] His book, *Driven to Innovate*, tells the stories of thirty-five Jewish mathematicians and physicists born in the 19th century, and he has also explored the links between mathematics and Asperger's syndrome. In 2007 he collaborated with Michael Fitzgerald on *The Mind of the Mathematician*.[34]

Richard Taylor

When Ioan James retired in 1995, the number theorist Richard Taylor became the Savilian Professor of Geometry. In the same year his joint paper with Andrew Wiles, completing the proof of Fermat's last theorem, was published in the *Annals of Mathematics*.

Born in Cambridge in 1962, Richard was even younger than Michael Atiyah had been on taking up the Savilian chair. He spent most of his childhood in Oxford before returning to Cambridge as an undergraduate at Clare College in 1980. Raised by parents who were a theoretical physicist and a piano teacher, he developed an interest in mathematics at an early age, and at Cambridge (where he was president of the Archimedeans) he became particularly attracted to number theory. He spent four years (1984–88) in Princeton completing his PhD degree under the supervision of Andrew Wiles, who

Richard Taylor.

had earned his doctorate (under the supervision of John Coates) while at Clare College in 1980. I myself overlapped with both Andrew Wiles and Richard Taylor while an undergraduate at Clare from 1977 to 1981.

After a postdoctoral year in France at the Institut des Hautes Etudes Scientifiques, Richard returned to Cambridge (where John Coates was now the Sadleirian Professor of Mathematics), becoming a Fellow of Clare College and assistant lecturer, then lecturer, and then Reader in the University. It was during these years that Andrew Wiles announced his proof of Fermat's last theorem during a lecture on 'Modular forms, elliptic curves and Galois representations' at the Isaac Newton Institute in June of 1993. Fermat's celebrated 'last theorem' from number theory had been formulated by the French mathematician Pierre Fermat just eighteen years after the founding of the Savilian chairs in 1619, and had remained unproved for 250 years.

Unfortunately, just three months after announcing his proof, Andrew Wiles realized that it contained a subtle, but serious, error. He tried to repair it, initially on his own, and then by enlisting Richard's help, and finally (a year later) he discovered how this could be done. In October 1994 he submitted two manuscripts to the *Annals of Mathematics* which were published in the following year after an extremely careful refereeing process.[35] The second of these was his joint work with Richard Taylor which provided a justification for the corrected step in the earlier paper.

Richard Taylor moved to Oxford as Savilian professor in 1995 after his marriage to Christine Chang in August of that year. He stayed in Oxford for just one year. His American wife, herself a mathematician and studying for a doctorate, was keen to return to the United States, and Richard accepted the offer of a professorship at Harvard

University. They did not move again until 2010 when Richard transferred to the Institute for Advanced Study in Princeton, first as a one-year visitor, and then as the Robert and Luisa Fernholz Professor of Mathematics. Since 2018 he has been a professor in the humanities and sciences at Stanford University in California.

Since the excitement caused by Fermat's last theorem, Richard Taylor has been a major contributor to spectacular developments in the theory of automorphic forms, including the proofs of a number of important conjectures; these have not, however, received the same media attention as did the proof of Fermat's last theorem. He was elected to a Fellowship of the Royal Society in 1995, and became a member of the US National Academy of Sciences twenty years later. He has received many honours and prizes, including the London Mathematical Society's Whitehead, Fermat, and Ostrowski Prizes, the American Mathematical Society's Cole Prize, and the Shaw Prize for Mathematics. In 2014 he was awarded one of the prestigious Breakthrough Prizes in Mathematics 'for numerous breakthrough results in the theory of automorphic forms, including the Taniyama–Weil conjecture, the local Langlands conjecture for general linear groups and the Sato–Tate conjecture'.

Nigel Hitchin

In 1997 another geometer became the Savilian Professor of Geometry. Nigel Hitchin's doctorate (completed in the early 1970s) had been supervised by Brian Steer and Michael Atiyah, and he moved from Oxford to the United States to work as Michael's research assistant at the Institute for Advanced Study in Princeton; he then visited the Courant Institute in New York before returning to Oxford. Like Michael's, his research interests are wide-ranging – including differential geometry, gauge theory, algebraic geometry, complex and quaternionic geometry, and mathematical physics – but they are clearly centred on geometry.

Nigel Hitchin describes his early life, student days, and career in the interview that follows this chapter. From 1974 he worked in Oxford, often in collaboration with Michael Atiyah, first as a research fellow and then as a tutorial fellow at St Catherine's College. In 1990 he moved to a professorship at Warwick University and in 1994 became Cambridge's Rouse Ball Professor of Mathematics, before returning to Oxford in 1997 to take up the Savilian chair which he held until his retirement in 2016. I was appointed as his successor.

Nigel's name has been attached to many significant mathematical advances over the last fifty years. Some of these involve joint work with Michael Atiyah, including the

Nigel Hitchin and Frances Kirwan.

'Atiyah–Hitchin–Singer theorem' (which uses the index theorem to compute the dimension of an instanton moduli space), the 'Atiyah–Hitchin metric' (on a moduli space of monopoles), and the 'ADHM construction of instantons' (with Atiyah, Vladimir Drinfeld, and Yuri Manin as co-authors). Other results have only Nigel's name attached, such as the 'Hitchin connection', the 'Hitchin equations', and the 'Hitchin system'. The 'Kobayashi–Hitchin correspondence' is fundamental in the study of the differential geometry of moduli spaces, while other important geometrical terms such as 'Higgs fields', 'Higgs bundles', and 'generalized geometries' were invented by Nigel and are inextricably linked with his name.

Nigel Hitchin was elected a Fellow of the Royal Society in 1991 and received the Royal Society's Sylvester Medal (named after James Joseph Sylvester) in 2000. He was President of the London Mathematical Society from 1994 to 1996, and was awarded its Senior Berwick Prize in 1990 and its Pólya Prize in 2002. In 2016 he received the Shaw Prize for Mathematical Sciences, an international prize that honours 'individuals who are currently active in their respective fields and who have recently achieved distinguished and significant advances, who have made outstanding contributions in academic and scientific research or applications, or who in other domains have achieved excellence'.

Nina Mae Foster's portrait of Frances Kirwan at Balliol College.

Conclusion

In an interview that took place in 2004 during the Abel Prize celebrations, Michael Atiyah observed:[36]

The passing of mathematics on to subsequent generations is essential for the future, and this is only possible if every generation of mathematicians understands what they are doing and distils it out in such a form that it is easily understood by the next generation. Many complicated things get simple when you have the right point of view. The first proof of something may be very complicated, but when you understand it well, you readdress it, and eventually you can present it in a way that makes it look much more understandable – and that's the way you pass it on to the next generation! . . . That has been done remarkably successfully for centuries.

Michael Atiyah died on 11 January 2019, at the beginning of the year that celebrated the 400th anniversary of the founding of the Savilian chair of geometry. He occupied that position for only six years, but his influence on his successors and on generations of mathematicians across the world was, and will continue to be, profound. Although short in physical stature, he was truly one of the giants of 20th-century mathematics.

Nigel Hitchin was Oxford's Savilian Professor of Geometry from 1997 to 2016. To celebrate his 70th birthday and honour his widespread contributions to mathematics, conferences were held in September 2016 in Aarhus, Oxford, and Madrid.

CHAPTER 8

Interview with Nigel Hitchin

MARK MCCARTNEY

Nigel Hitchin has spent most of his academic life at Oxford. Commencing as an undergraduate at Jesus College in 1965, he continued at England's oldest university for his doctoral studies and held appointments as researcher and college fellow before being appointed Savilian Professor of Geometry in 1997. While the centre of gravity of his life has been in Oxford, his career has been interspersed with significant periods at Princeton, Stony Brook, Warwick, and Cambridge. In this interview he reflects on a career in mathematics which has spanned more than half a century.

Early life

You were born in Holbrook, Derbyshire on 2 August 1946. Tell me about your early home life and school education.

I grew up first in Duffield, just north of Derby, attending the village school there. My father was an industrial chemist, a plant manager at British Celanese, a large factory on the other side of Derby. He had worked there since he left school, but only after he married did my mother persuade him to take an external London degree – after school he was more interested in motor bikes than further education. My mother left school at 14 to help her widowed father who ran a bakery. After her marriage she stayed as a housewife. My brother was four years older than me. He taught me to read quite early on, and for

Mark Mccartney, *Interview with Nigel Hitchin*. In: *Oxford's Savilian Professors of Geometry.*
Edited by Robin Wilson, Oxford University Press. © Oxford University Press (2022). DOI: 10.1093/oso/9780198869030.003.0008

many years I followed his interests. He used to design and create sophisticated Meccano constructions with complicated gearing: I could never do that. He was also in the Cadet Force at his grammar school, and I followed his enthusiasm for aviation. But my secondary school was a different one, and gradually I began to get interested in other things.

Were you particularly attracted to mathematics and the sciences at school? What sort of mathematics and science curriculum did you experience?

The Ecclesbourne Grammar School in Duffield was a completely new co-educational school, and in 1957 I was one of the first pupils: there were 74 of us 11-plus entrants making two forms, and a handful of 13-plus boys to act as prefects and keep us under control. Small numbers meant individual attention, but also few teachers, so that, for the first year at least, mathematics was taught by the French master. The school was modern, but the headmaster had traditional values and liked to model his school to some extent on a public school. We had to play rugby (not soccer), learn Latin, and establish traditions even though the school was new. But he had a degree in mathematics and he kept a watchful eye on budding mathematicians, especially in the sixth form. After a year we did have a dedicated mathematics teacher, Norman Else, who was keen to answer questions about how and why. There were always three of us vying for top marks in mathematics or physics.

For several years I tended to think of engineering as a career, but after missing the first few weeks of the initial sixth-form term because of a serious appendix operation I dropped one subject and focused on mathematics and physics for A-levels. It then looked as though I would be applying to do mathematics at university.

The headmaster was naturally keen on getting some of his first cohort to Oxbridge, and we three did go, two of them straight after A-levels, but I was persuaded to stay on for the Scholarship Examination. I was offered a place at Jesus College, Oxford, in October 1964 after A-levels, but the tutors again advised me to take the examination. Why Jesus? The headmaster had been offered a place there in the 1930s but could not afford to take it up, and I realized then how vicariously he had been monitoring my progress. In December of that year I got a telegram offering me an Open Scholarship, and I left school behind.

Before university you spent 8–9 months working in the Engineering Computing Department at Rolls Royce in Derby. How did this role come about, and what sort of work did it involve?

It was common at that time for students in my position to get some experience vaguely related to future study at university; there was no opportunity for going around the world in a gap year unless you were really well off. Rolls Royce was one of the other big employers in Derby and the contact was made through the headmaster. I went for an interview and answered wrongly a question about coordinate dependence of random points in the plane, but nevertheless got a job as a trainee programmer in the Engineering Computing Department for the princely wage of £7 12s per week.

I learned Fortran and I was given various jobs. The skills needed were organizational and combinatorial: for example, finding a pattern of mixing nozzle guide vanes in a jet engine to avoid the natural frequency of the turbines, without using months of computer time: I knew nothing of Fourier analysis at the time, but eventually realized why I was doing this.

Some of the other programmers would ask me mathematical questions about their work. I remember one about the 'stiffness matrix' – I was asked why if the sum of the rows of a matrix was 1, the same was true of its powers. I knew virtually nothing about matrices (it was not in any syllabus at school), but I did know how to multiply a vector by a matrix and choosing the obvious vector gave the answer. So by osmosis I absorbed something about mathematics there.

I also absorbed too much pie and chips in the canteen, and put on a lot of weight! I returned for a couple of periods of vacation work and they told me that there would always be a job for me if I wanted one after graduation. When things got difficult as a research student I considered it, but by then I really wanted to widen my horizons and not return to Derby.

A student at Oxford

In 1965 you entered Jesus College, Oxford, to study mathematics. What were Oxford and college life like in the 1960s?

Coming for interview in December 1964, Oxford was a rather forbidding place. The black rain-soaked walls were more reminiscent of Jude the Obscure than the way they look to-day – and there were no signs outside telling you which college was which. However, when I came up in October 1965, I was given a bedroom and sitting room in a staircase overlooking Ship Street, reasonably well heated and comfortable. Some of my friends were in the older rooms and were to be found huddled over electric fires in the winter months. There was still a traditional air about the place, with tea, buttered toast, and gentlemen's relish in the Junior Common Room every day. There was a dominant Welsh presence in the college, and the common room was packed when there was a rugby match.

Nigel Hitchin, ready to take his Oxford Finals examinations in 1968.

There were eight mathematicians at Jesus in that year, and three of us had academic careers. Lyn Thomas, who died in 2016, became Professor of Management Science in Southampton and sometime President of the Operational Research Society, and Gareth Jones became professor in the mathematics department there. There were the Moderations exams at the end of the first year and then nothing till Finals, so the summer of the second year was a relaxing interval. Whenever I hear 'A whiter shade of pale', it reminds me of that time.

Do you have memories of any particular lecturers or tutors during your time as an undergraduate?

On the first day of Mods lectures we encountered the top brass: algebra was taught by the Savilian Professor of Geometry, Michael Atiyah, applied maths by the Rouse Ball Professor, Charles Coulson, and analysis by John Hammersley. Hammersley was an expert in percolation theory, but also an opponent of the 'New Maths', so his question sheets were all hard problems with links to his own applied work. He is renowned for having set in 1966 the most difficult set of compulsory papers in living memory. To some students his problem sheets were a challenge to be attacked – but most gave up and asked their tutors. The Faculty did not repeat the experiment.

Within the College the tutors were Edward Thompson in pure mathematics and Christopher Bradley in applied. Thompson's attitude was one of bringing your unsolved problems to him, and in a cloud of pipe smoke he would outline a solution. He would sometimes hint at his experience in (what we learned later was) Bletchley Park, but was increasingly involved with College affairs. Bradley was happy to give more individual attention. I liked his approach to applied mathematics, which was very clean, and he introduced me, for example, to tensors. He was happy to give up time to explain things outside the syllabus.

In 1969, after graduating with your BA, you moved to Wolfson College, Oxford, to study for a DPhil degree, with your supervision starting with Brian Steer and then moving to Michael Atiyah. What were Steer and Atiyah's styles of supervision? Did you find it easy to adapt to working on large-scale research problems?

In 1968 I stayed on as a graduate student in Jesus for a year, and Brian Steer at Hertford College was assigned to me as supervisor: he had given me tutorials in topology and differential geometry in my third year as an undergraduate. For my first year I found it difficult to settle down to the methods of doing research. There was an examination at the end of that year leading to a Diploma in Advanced Mathematics, so there were some courses to be attended, but finding a research problem that I liked, or could make headway in, was difficult.

Brian tried me first with dynamical systems. This was supposed to be an exciting subject – Berkeley was the epicentre, and I attended a summer school at Warwick to learn more. But in the end that did not appeal. He then tried me with a problem in K-theory, so I read Atiyah's book but could not link it to the problem at hand. I was getting a bit depressed when he showed me a short paper by Lichnerowicz about the Dirac operator, and with this I made some progress. I studied the papers on the Atiyah–Singer index theorem which were just coming out, and wrote a minor extension of the Lichnerowicz result for the required dissertation. It was enough to give me confidence to proceed.

At the time Wolfson College was trying to build up a graduate population with small scholarships, and I obtained one and moved college in 1969 – not to a college building as it exists now, but to a house on the Banbury Road. It was a very positive experience, with meals in close proximity to like-minded graduate students, postdocs, and lecturers – an environment more conducive to the pattern of life as a graduate student.

Within the Mathematical Institute I shared an office with a student of Michael Atiyah (and at one stage also with Don Zagier before he moved to Bonn). It seemed that he was doing more interesting stuff than me, and Atiyah's Monday seminar was the highlight of the week. As my research progressed, Brian explained some of it to Atiyah and, since it resonated with something he and Raoul Bott were working on, he agreed to supervise me for a term while Brian was away on sabbatical. Soon after, Atiyah moved to the Institute for Advanced Study in Princeton, but came back to Oxford for Trinity Term, so the supervisions continued, though now not officially under the University's auspices. Supervisions were very lively, with Atiyah offering suggestions, sketching out the background on the blackboard, informing me about the essentials of a subject and not asking me to get a book to read about it. It was difficult to recall everything and to try and write it down afterwards, but it was an amazing experience.

Post-doctoral work

Between 1971 and 1974 you worked as Atiyah's research assistant at the Institute for Advanced Study (IAS) at Princeton, and then spent a year at New York University (NYU). Were there significant differences between the scholarly environments in the IAS and NYU, compared to Oxford?

Moving to Princeton as Atiyah's research assistant was an eye-opener for me. It wasn't just being in America, but the fact that you were surrounded by research mathematicians, young and old, who knew much more than yourself but had time to talk – especially more senior people who were not hampered by teaching duties. It was a worldwide centre of mathematics and new ideas often made themselves known there first. My duties as an assistant were minimal, and to all intents and purposes I was like a normal visitor. I had invitations to speak in other universities on the east coast and learned much more about the mathematical community. I also met my wife there – she was visiting her cousin, another Institute visitor.

In 1973 I got married, and we moved to New York where my wife was still studying. The year at the Courant Institute at NYU was very different, as was life in general: Derby, Oxford, and Princeton were villages compared with New York. I missed the opportunity to clarify my thoughts by going for a long walk – there are limits to walking around Washington Square, and the alternative meant stopping for traffic at each street corner which disturbs your thoughts. I taught a couple of courses, some in the evening which was new for me. At the time, I didn't think that my research advanced so much, but in retrospect the things I quietly studied there became relevant later.

In 1974 you returned to Oxford to a sequence of two Science Research Council research positions which lasted until 1979. This meant that from the start of your doctoral degree you had ten years of uninterrupted research. What were the most significant pieces of research work you completed over that decade?

I actually finished my thesis in my first year at Princeton, and it was there that I observed that a rather subtle mod 2 index could be used, together with some established results in topology, to get information about positivity of scalar curvature with some unexpected results on exotic spheres. I never went back to that problem, but in recent years with more sophisticated methods there is a growing community of researchers.

From 1975 in Oxford I had been learning about Penrose's twistor theory, and then when Isadore Singer came for a term and introduced us to the Yang–Mills equations, many things I had been thinking about fell into place, and with Atiyah we gave a complete construction of self-dual solutions to these equations. This was the ADHM (Atiyah–Hitchin–Drinfeld–Manin) construction of instantons.[1]

Around the same time I was in contact with Stephen Hawking and Gary Gibbons in Cambridge about gravitational instantons, and again using twistor theory showed how the theory of Kleinian singularities and Dynkin diagrams linked up with what they were doing.

College fellow

Between 1979 and 1990 you were Fellow and Tutor in Mathematics at St Catherine's College, Oxford. How did you adapt to the change from full-time research freedom to significant teaching duties? Did you notice any significant changes in the Oxford mathematics curriculum in the 1980s from the one that you had experienced in the 1960s?

St Catherine's was kind to me in interpreting the long hours of tuition required for the position of CUF (Common University Fund) Lecturer – the balance between University and College teaching weighed more heavily with the College – but the duties did impinge on my research time. On the other hand, having time available does not mean that ideas come any quicker – in fact, sometimes you get them in a meeting or listening to a lecture. I worked in the Mathematical Institute in the mornings and often, since Michael Atiyah was a professorial fellow at the same college, we would walk together across the University Parks from the Institute for lunch and exchange ideas then. The constraints I felt were more that teaching time took away the opportunity to go to seminars or conferences, both in Oxford and elsewhere.

At any rate my research proceeded well, despite any such constraints, and I developed the field of magnetic monopoles and the work on Higgs bundles which later on became quite influential.

Mathematicians in animated conversation: Nigel Hitchin with Michael Atiyah and Shing-Tung Yau at a Durham symposium in 1982.

As far as the courses I was concerned with teaching, there was not much difference between the 1980s and the 1960s: I think that the differences would have been more noticeable in the third-year specialist subjects. Together with Graeme Segal, also at St Catherine's, we introduced new courses in geometry which were lacking when I was a student.

During the academic year 1983–84 you visited the State University of New York at Stony Brook, with the opportunity to stay there permanently. What made you return to Oxford? Can you imagine an 'alternative life' in which the rest of your career could have been in the United States?

In 1982 I had given a talk in Stony Brook, and over a lobster dinner they approached me about a senior position there. I thought about it, and then took a sabbatical to try it out. I had a young family then, money was tight, and when I thought about it, maybe more research time could be useful. We rented a large house, made friends, and enjoyed ourselves there. I also had a very productive cooperation with physicists which some years later appeared as a much-cited paper.[2] I had to teach both undergraduates and graduates but, apart from aggressive pre-med students who were unhappy with their grades, this was fine. Then the time came to make a decision to stay or not.

From a domestic point of view, had I stayed I would have had a much better salary than in Oxford, and lived in a better house, but would the rest of our lives have been better or the same? In Oxford, in the meantime, the college had actually approved a buy-out of some of my teaching time, paid for by Michael Atiyah's research grant – this was unusual in those days. I had several meetings with the Stony Brook chairman, and also one with C. N. Yang of Yang–Mills fame, trying to persuade me to stay, but in the end we came back.

What would my alternative life have been? Would I have moved on from Stony Brook? I have returned there frequently in the last ten years as a trustee for the Simons Center for Geometry and Physics, but if I compare my life as a faculty member there or here, I think I am happier here.

The 1990s: a decade of three chairs

In 1990 you were appointed professor of mathematics at Warwick, and then in 1994 you moved again, this time to Cambridge as Rouse Ball Professor of Mathematics, and then finally you moved back to Oxford to the Savilian chair of geometry in 1997. What prompted these moves? What were the similarities and differences between Oxford, Cambridge, and Warwick?

In 1990 Michael Atiyah left Oxford to become Master of Trinity College, Cambridge, and the prospect loomed of returning to full-time teaching. Over the years I had been approached about professorships in other universities, but never really followed them up as I was happy in Oxford – but at the age of 43 it did seem a time to move on. Christopher Zeeman had just retired from Warwick to become the head of an Oxford college, and his chair was vacant, so I applied. After some internal discussion in Warwick, the details of which I never knew, two professorships were created, the other going to Alan Newell, an applied mathematician.

I found Warwick a very friendly place. I knew several people there, including the geometer John Rawnsley, with whom I had shared houses as a student in Oxford, and the analyst Jim Eells. I used to go up to Warwick to attend seminars as a graduate student when they had a special year in algebraic geometry, and I got to know Jim then; I also met him later in Princeton where he encouraged my work.

To my surprise, I was not free of tutorials – all professors and lecturers were assigned students to see every week. As a professor I was more involved with committees and

Among his many honours, Nigel Hitchin has been awarded honorary doctorates from the Universities of Bath (2003) and (pictured here) Warwick (2014).

outside issues, for example on the Mathematics Committee of the Science and Engineering Research Council. Graduate students were assigned to me, rather than my choosing them. While at Warwick I was elected to a Fellowship of the Royal Society in 1991 and then, as is the custom, was put on committees there. Emerging from one of these, the Cambridge head of department took me aside for a coffee and directed me towards the Rouse Ball chair which had become vacant on John Thompson's retirement.

I knew that Ioan James would retire from Oxford's Savilian chair in 1995, then just a year or so ahead, and that would have been my target, but you don't as an English mathematician turn down the chance of being a professor in Cambridge – a bird in the hand, and all that. So I applied and was appointed.

I actually knew little about Cambridge before going there. Named professorships are not associated with colleges, but after the experience of Warwick I rather looked forward to returning to a college environment. I was lucky to be admitted as a professorial fellow at Caius College, sharing a room with Stephen Hawking. On my admission I was welcomed as 'having overcome the misfortune of an Oxford education' and the college treated me well. Although it was not part of my duties, I volunteered to supervise undergraduates there, which I actually found rewarding because they were really very good.

The Department of Pure Mathematics was then in a low-ceilinged building, previously used as a book depository by the University Press. Fundraising was just beginning for a new centre, but the building was quite uncomfortable compared with the Oxford and Warwick institutes. One redeeming feature for me was the way in which mathematicians and physicists came together. The geometry seminar which I was running was held in the applied mathematics department across the road, so that Hawking could attend. Together with Graeme Segal, who had moved to Cambridge's Lowndean chair in 1990, the seminar focused on attempting to explain new pieces of mathematics coming from string theory or other branches of theoretical physics, with external speakers roped in to tell us more.

There was of course the routine work of examinations. When I was appointed as chairman of some exam board, the information came round that 'the procedures will be carried out in the usual way'. I did feel sometimes that there was an underlying assumption that if you were on the faculty in Cambridge, you must have been there all your life.

After two years I was informed that I was next in line to be Head of Department. But soon afterwards I received a phone call from Oxford concerning the Savilian chair, as Richard Taylor had departed for Harvard. A few phone calls revealed that a job opportunity was possible in the John Radcliffe Hospital for my wife, and after some deliberation I applied and was offered the position. Cambridge had to find someone else to be Head, which was a bit of a shock for the person chosen.

Savilian Professor of Geometry

Not all holders of the Savilian chair have been geometers, but it is a term that does apply to your research. You have spent your career working at the very deep interface between mathematics and physics. In what ways do you see geometrical thinking as a key aspect of modern mathematical physics?

Geometrical objects in mathematics carry with them an aura of mental concepts – visual, tactile analogues of the physical world around us – which help us to open up avenues of exploration using whatever mathematical tools are at hand. Physicists like to calculate, but that requires a condensation from a concept to a formula, and geometry is a good way to hold that concept – or even better, to draw parallels from other areas to aid the mathematical formulation. Of course, general relativity and Yang–Mills theory are the classical examples of geometry coming to the aid of physics.

Do you have any personal memories of your recent predecessors in the Savilian chair of geometry?

From what I have just said, clearly Michael Atiyah had a huge influence on my research, and his successor, Ioan James, was a constant presence in my time in Oxford until I left for Warwick, running seminars and guiding DPhil students towards me, including Simon Donaldson on the basis of a letter commenting on 'the best in Cambridge for ten years'. I sometimes wonder who the other student ten years earlier had been.

You have been elected to Fellowship of the Royal Society, been President of the London Mathematical Society, and been presented with various awards. Can you tell me about some of these?

It is always gratifying to obtain recognition for your efforts. At the International Congress of Mathematicians in Helsinki in 1978 I met up with mathematicians I had known in Princeton, and I was pleased to hear that the ADHM work had spread widely. I was a speaker at the next ICM in Warsaw, held in 1983 because of martial law in 1982.

The first prize that I received was the LMS Whitehead Prize in 1981, which I always remembered when listening to music as it enabled me to replace a second-hand record player bought on my return from America! The Shaw Prize which was awarded just before my retirement was a huge surprise, but one which confirmed the international status of my work.

The election to the Royal Society led rather naturally to my being approached to be President of the London Mathematical Society. I was at Warwick at the time, and I overheard a comment, 'Thank goodness it is not Oxbridge again!', but soon afterwards I moved to Cambridge and it was from there that I went to Council meetings. It was a period when the limitations of Burlington House were becoming more apparent, and money was put aside to prepare for a move some time in the future. My successor, John Ball, took the bull by the horns and bought De Morgan House which changed the face of the Society.

My LMS, Royal Society, and SRC experiences meant that I had a much wider perspective of UK mathematics and this led to my participation at increasing levels in the Research Assessment Exercises. In the final 2008 exercise the main panel had mathematicians, statisticians, and computer scientists under the same umbrella, which sometimes formed an uneasy alliance.

(*Left*) In September 2016 Nigel Hitchin received the Shaw Prize for Mathematical Sciences from Chun-Ying Leung, Chief Executive of Hong Kong, 'for his far-reaching contributions to geometry, representation theory, and theoretical physics'.
(*Right*) Nigel Hitchin giving his Shaw Lecture at the Hong Kong University of Science and Technology (HKUST).

Your career in mathematics has lasted over fifty years. Who do you consider to be the most significant mathematicians and physicists that you have met over your career?

In different ways, the mathematicians in Michael Atiyah's orbit all influenced my work. Singer, in particular, expressed an interest in what I was doing at several stages of my career: he had more of a differential-geometric and analytic background than Atiyah. When I was a postdoc at NYU, I remember going up to Harvard to give a talk one cold December day, and finding to my horror that the audience consisted of just Atiyah, Isadore Singer, Bertram Kostant, and Irving Segal. The talk was about the Dirac operator and the Szegő kernel. I didn't know how much I had communicated, but a few days later Singer sent me several pages showing a more sophisticated approach.

And it was Singer who brought the instanton problem to Oxford. He would run a joint mathematics–physics seminar at MIT, so he was really the first facilitator in my interaction with physics. Often the information would come via Michael Atiyah, who was always crossing the Atlantic to learn new things. Once string theory became popular there were all sorts of challenges for mathematicians, and I relied on physicists with the patience and ability to distil from their problems questions which a mathematician could at least absorb. Edward Witten is the most obvious example.

Over the next fifty years, what do you think are likely to be the most significant changes in mathematics – both in terms of which research directions are pursued, and how research is likely to be carried out?

A mathematician with the modern tools of his trade: paper, pen, and laptop.

The Internet has already changed enormously the way that we do research – if you get an inkling that some topic might be relevant to your current problem, then Wikipedia, math.stackexchange, MathSciNet, etc., take you pretty quickly to the current information. It is often not enough, and asking someone might be better, but then you can email them too. How this will evolve in the future is difficult to say. The volume of published material keeps on increasing, and mathematics may become more fragmented as a consequence. These trends present challenges to find a better way of interacting with the computer.

One of the changes in my area is the increasing use of categorical language as the framework in which results are expressed. I often find this difficult, but younger mathematicians not so. And just as I found the concept of a manifold easy to absorb as a student, in the writings of Élie Cartan and William Hodge in the 1930s, it is more obscure – so this is part of the onward progress of the subject. However, when it comes to higher category theory it is difficult to find the right words, geometrical or otherwise, to describe what is going on, and here I think the role of the computer will become more important – not just proof-checking (for example, treating the higher-level analogues of commutative diagrams), but providing a practical interface. If the computer understands the concepts and records our internet searches, then perhaps it will make useful suggestions on how to proceed to a proof.

FURTHER READING, NOTES, AND REFERENCES

This section contains suggestions for further reading for each chapter, and detailed *notes and references* as indicated within the chapters.

Works cited in more than one chapter include:

John Fauvel, Raymond Flood, and Robin Wilson (eds), *Oxford Figures: Eight Centuries of the Mathematical Sciences*, Oxford University Press (2nd edn, 2013). [*Oxford Figures*]

R. T. Gunther's *Early Science in Oxford*, 15 vols, Oxford University Press (1920–67). [*Gunther*]

Reading Mathematics in Early Modern Europe: Studies in the Production, Collection, and Use of Mathematical Books (ed. Philip Beeley, Yelda Nasifoglu, and Benjamin Wardhaugh), Routledge (2020). [*Reading Mathematics*]

Also useful are the various volumes of
The History of the University of Oxford, Clarendon Press (1984–2000).

In general, for books published before 1900 only the places of publication are given, and for books published thereafter only the publishers are given.

In these notes, *Phil. Trans.* refers to the *Philosophical Transactions of the Royal Society*.

CHAPTER 1

Further reading

The following list of references may be supplemented with the article:

William Poole, 'The Origin and Development of the Savilian Library', in *Reading Mathematics*.

For a more general account, see also

William Poole and Christopher Skelton-Foord, *Geometry and Astronomy in New College, Oxford: On the Quatercentenary of the Savilian Professorships, 1619–2019*, New College, Oxford (2019).

Notes and references

1. There is no adequate biography of Henry Savile but attempts commenced with the funeral oration by Thomas Goffe prefaced to the *Ultima Linea Savilii* (Oxford, 1622), a memorial volume including verse by both Savile's first professors.

2. See Ronald G. Cant, *The College of St Salvator: Its Foundation and Development, including a Selection of Documents*, University of St Andrews (1950), 172–4.

3. Andrew Dalzel, *History of the University of Edinburgh from its Foundation*, Vol. II (Edinburgh, 1862), 74, 336–44.

4. P. J. Anderson (ed.), *Fasti Academiae Mariscallanae Aberdonensis*, Vol. I (Aberdeen, 1889–98), 133, and P. D. Omodeo (ed.), *Duncan Liddel (1561–1613): Networks of Polymathy and the Northern European Renaissance*, Brill (2016), 195–7.

5. G. R. M. Ward (ed.), *Oxford University Statutes*, Vol. I (London, 1845–51), 272.

6. As note 5.

7. William Burton, *In Viri Doctissimi, Clarissimi, Optimi, Thomae Alleni ... Orationes Binæ* (London, 1632), 5.

8. [Seth Ward], *Vindiciae Academiarum* (Oxford, 1654), 28.

9. See Robert Goulding, 'Henry Savile reads his Euclid', *For the Sake of Learning: Essays on Honor of Anthony Grafton*, Vol. II (ed. A. Blair and A.-S. Goeing), Brill (2016), 780–97, and, more generally, Goulding's *Defending Hypatia: Ramus, Savile and the Renaissance Rediscovery of Mathematical History*, Springer (2010).

10. See Bodleian Library, MS Savile 10, and Mark L. Sosower, 'Greek manuscripts acquired by Henry and Thomas Savile in Padua', *Bodleian Library Record* 19 (2006), 157–84.

11. See Mordechai Feingold, 'Patrons and professors', *The 'Arabick' Interest of the Natural Philosophers in Seventeenth-Century England* (ed. G. A. Russell), Brill (1994), 109–27, on pp. 116–18.

12. Thomas Smith (ed.), *Viri Clarissimi Gulielmi Camdeni ... Epistolae* (London, 1691), 313–15.

13. See Philip Beeley, '"A designe Inchoate": Edward Bernard's planned edition of Euclid and its scholarly afterlife in late seventeenth-century Oxford', in *Reading Mathematics*.

14. See Ward (note 5), pp. 273–4.

15. There are no surviving portraits of Henry Briggs. The older biobibliographical accounts are John Aubrey, *Brief Lives* (ed. K. Bennett), Oxford University Press (2018), 146–9; Anthony Wood, *Athenae Oxonienses*, Vol. II (ed. P. Bliss) (London, 1815), 491–3; Thomas Smith, 'Vita Briggii', in 'Commentariolus de vita et studiis . . . D. Henrici Brigii', *Vitae Quorundam Eruditissimorum et Illustrium Virorum* (London, 1707); John Ward, *The Lives of the Professors of Gresham College* (London, 1740), 120–9. Smith was greatly aided by notes from the Cambridge antiquary Thomas Baker (Bodleian, MS Smith 47, fols. 23, 47–51). Some additions are provided by D. M. Hallowes, 'Henry Briggs, mathematician', *Halifax Antiquarian Society* (1961), 79–92, and the *Oxford Dictionary of National Biography*.

    Briggs also plays a significant role in Nicholas Tyacke, 'Science and religion at Oxford before the Civil War', *Puritans and Revolutionaries* (ed. D. H. Pennington and K. Thomas), Oxford University Press (1978), 73–93, and Mordechai Feingold's indispensable *The Mathematicians' Apprenticeship: Science, Universities and Society in England, 1560–1640*, Cambridge University Press (1984).

16. Henry Savile, *Praelectiones Tresdecim in Principium Elementorum Euclidis* (Oxford, 1621), 260.

17. Aubrey (note 15), p. 265.

18. Feingold (note 15), pp. 68–9.

19. Feingold (note 15), p. 79.

20. Feingold (note 15), pp. 50–1. The lectures and their provenance were recorded by Wood (note 14), p. 492, identified as British Library, MS Harl. 6796, fols. 81–54 (item 17) by Feingold, and endorsed by Apt with the comment, however, that at this stage, despite a mention of Copernicus, they show Briggs to have been a conventional Ptolemaic in his cosmology.

21. J. E. B. Mayor, *Early Statutes of the College of St. John the Evangelist*, Cambridge (1859), 328; Richard Rex in P. Linehan (ed.), *St John's College: A History*, Boydell (2011), 18.

    The St John's rubric follows closely the *c*.1500 statute for the University lectureship, usually held, as elsewhere, by regent masters: see Paul Lawrence Rose, 'Erasmians and mathematicians at Cambridge in the early sixteenth century', *Sixteenth Century Journal* 8 (1977), 46–59.

    For Briggs's election see in the College archives the register SJAR6/1/1/1 (*examinator*, 7 July 1592; *lector Medicinae pro Doctore Linacre*, 8 September 1592), and for the corresponding rental SJAR/3/2/4/2 (10s a quarter for the examinatores, but £3 a term for the *lector medicinae*).

22. Ward (note 15), p. 126, from Gataker's testimony in his *Vindication of the Annotations . . . Jer. 10.2* (London, 1653), 87. The same passage shows that when

Gataker later consulted Briggs at Gresham College on judicial astrology, Briggs reassured him it was 'a meer System of groundlesse conceits'.

For more anecdotes of Briggs's declarations against astrology, see John Geree, *Astrologo-Mastix* (London, 1646), 14–15, and William Rowland, *Judiciall Astrologie Judicially Condemned* (London, 1651), 63–6.

23. Anthony Wood, *History and Antiquities of the University of Oxford* (ed. J. Gutch), Vol. II (1796–98), 1, 334–5.

24. Smith (note 15), p. 2.

25. Letter from Henry Briggs to Samuel Ward, 6 August 1621, Bodleian, MS Tanner 73, fol. 68.

26. Robin J. Wilson, 'The Gresham Professors of Geometry Part 1: the first one hundred years', *BSHM Bulletin: Journal of the British Society for the History of Mathematics* 32 (2017), 125–35, on pp. 129–30.

27. Ward (note 15), p. viii.

28. These tables can be found in Thomas Blundeville, *Theoriques of the Seven Planets* (London, 1602), sig. [Ppiv]v, and Edward Wright, *Certaine Errors in Navigation*, London (1610), various tables, attributed on [A5]r–v. It is notable that Wright in his preface to the reader (¶[7]r) reproduces in English translation part of a letter from the cartographer Jodocus Hondius to Briggs, excusing his conduct for publishing material from Wright without consent.

29. Smith (note 15), p. 3; *The Correspondence of James Ussher: 1600–1656*, Vol. I (ed. E. Boran), Irish Manuscripts Commission (2015), 73–4, 108–9.

30. Adam Jared Apt, 'The Reception of Kepler's Astronomy in England: 1596–1650' (D. Phil. thesis, Oxford, 1982), esp. 176–87, and see Feingold (note 15), pp. 138–9.

31. *Philologiæ ΑΝΑΚΑΛΥΠΤΗΡΙΟΝ* (Oxford, 1652), 49; Aubrey (note 15), p. 149. He also wrote a Latin verse panegyric on logarithms in the same volume, on pp. 54–6. Both Briggs and Bainbridge are included by Sir Edward Sherburne in his influential 'Catalogue of astronomers ancient and modern', appended to his book *The Sphere of Marcus Manilius* (London, 1675), 86, 96.

32. Narratives of Briggs's encounter and collaboration with Napier are legion, but they start with Briggs's own account in his preface to the *Arithmetica Logarithmica* (London, 1624), sig. A3r–v, soon followed by (e.g.) Edmund Wingate, *Arithmetique Made Easy* (London, 1630), sigs. A1r–A3v.

The account of their first meeting appears in William Lilly's *Mr Lilly's History of his Life and Times* (London, 1715), 105–6. Briggs is clear that he lectured on improvements to Napier before contacting him, but Napier had promised in the 'Admonitio' at the end of some copies of his work, if it proved acceptable, to publish 'an explanation or method for either emending this rule, or for making afresh a more correct one' ('rationem ac methodum aut hunc canonem emendandi, aut emendatiorem de novo condendi'), *Mirifici Logarithmorum Canonis Descriptio* (Edinburgh, 1614), sig. m2r. See also Charles Hutton, *Mathematical Tables* (London, 1785), 20–41, and for more technical discussions of Briggs and logarithms, see Ian Bruce, 'The Agony and the Ecstasy: The development of logarithms by Henry Briggs', *Mathematical Gazette* 86/506 (2002), 216–27.

33. Katherine Neal, *From Discrete to Continuous: The Broadening of Number Concepts in Early Modern England*, Springer (2002), 80–114; this is the authoritative account.

34. John Napier (trans. Edward Wright), *Admirable Table of Logarithms* (London, 1616), sigs. [I7]r–[K2]v with table. As neither author nor translator was alive by this date, Briggs probably saw this through the press.

35. The *Short Title Catalogue* reported that 'both' surviving copies (at Christ Church and Balliol) were bound with Gunter, but English Short Title Catalogue was able to add exemplars from the British Library and Columbia, and to report from Balliol that its copy was not in fact bound in this way. The Christ Church volume (d.8.100) contains a third mathematical item, but the Gunter tract, standing first, was given by the author.

36. Briggs's bequeathed copy is now Bodleian, Savile N 3, and for the extra chiliad, see Noel Malcolm and Jacqueline Stedall, *John Pell (1611–1685) and his Correspondence with Sir Charles Cavendish: The Mental World of an Early Modern Mathematician*, Oxford University Press (2005), 303, and fn. 153 for details. For Briggs's bequests, see below.

37. For Vlacq's mathematical publications, see Leo Miller, 'Milton and Vlacq', *Papers of the Bibliographical Society of America* 73 (1979), 145–207, on pp. 154–64. The *Trigonometria Britannica* is in two books, the first by Briggs and the second by Gellibrand. For a study of Briggs's portion, see Ian Bruce, 'Henry Briggs: The Trigonometria Britannica', *Mathematical Gazette* 88/ 513 (2004), 457–74.

38. Susan M. Kingsbury (ed.), *Records of the Virginia Company*, 4 vols, United States Government Printing Office (1906–35), Vol. I, pp. 213–386 *passim*. Briggs was replaced on the court in May 1621, as he was 'nowe gon to Oxford to abyde' (Vol. I, 467). He is listed as a major shareholder in January 1622 (Vol. III, 592), and later returned to attend a court meeting in April 1623 (Vol. II, 341). See also Neal (note 32), pp. 104–6.

39. *Purchas His Pilgrimes*, Vol. III (1625), 852. The first edition is rare, and when Briggs's first biographer, Thomas Smith, was sent a copy by Hans Sloane, he replied with thanks but initially found it hard to believe that it was genuinely by Briggs (MS Smith 65, p. 221, Smith to Hans Sloane, 23 November 1704).

40. See the broadsheet dated 11 January 1629[/30] (STC 24000.5).

41. Feingold (note 15), p. 141.

42. Johann Kemke, *Patricius Junius, Patrick Young, Bibliothekar … Mitteilungen aus seinem Briefwechsel* (Leipzig, 1898), 55, 62 (letters from Patrick Young to Lucas Holstenius, May 1625, and reply of 28 May 1628). See also correspondence from Briggs to Holstenius, 4° Cal. Octob. (= 28 September) 1626 (Bodleian, MS D'Orville 52, fol. 134).

43. Smith (note 15), p. 14.

44. William Poole, 'The evolution of George Hakewill's *Apologie or Declaration of the Power and Providence of God*, 1627–37: Academic contexts, and some new angles from manuscripts', *Electronic British Library Journal* (2010), article 10, 8–9.

45. George Hakewill, *An Apologie or Declaration for the Power and Providence of God* (Oxford/London, 1635), 301–2.

46. Hakewill (note 45), pp. 107–8.

47. Bodleian, MS Bodley 313, letter of 1 June 1626; William H. Sherman, *John Dee: The Politics of Reading and Writing in the English Renaissance*, Massachusetts University Press (1995), 117–18. The committee comprised Savile, Dee's ward Thomas Digges, and Savile's Merton colleague John Chamber.

48. See Feingold (note 15), pp. 65–6, from the diary of Thomas Crossfield of Queen's College; Malcolm and Stedall (note 36), pp. 19, 254–5.

49. It was printed by William Jones, also the printer of Gunter's *Canon Triangulorum* in the same year, and probably therefore of Briggs's *Logarithmorum Chilias Prima*. Briggs later evidently hoped that the Parisian publisher Morel would take on the complete edition, but nothing came of this (Bodleian, MS D'Orville 52, fol. 134r (Briggs to Holsten, 4° Cal Octob. [= 27 September] 1626).

50. London, National Archives, PROB 11/159/216. See also Poole, 'Savilian Library'.

51. The older biobibliographical authorities are Wood (note 15), Vol. III, pp. 67–9; Smith (note 15), and more recently, see Tyacke (note 15) and especially G. J. Toomer, *Eastern Wisedome and Learning: The Study of Arabic in Seventeenth-Century England*, Oxford University Press (1996), 72–5.

52. A few copies of the printing are dated 1618 and were published by Henry Featherstone, but the work was entered into the Stationers' Register on 1 January 1619, and the bulk of the edition, from the same printer, is dated to that year, and was published by John Parker.
   For Savile's ban on astrology, see Ward (note 6), p. 274, and for more on anti-astrological works in the Savilian manuscripts, see Poole, 'Savilian Library'. Bainbridge nevertheless maintained a friendly correspondence with Sir Christopher Heydon, the chief writer in support of judicial astrology.

53. It was never published, but a text may be found in Trinity College, Dublin (MS 382, fols. 24r–29r). This general lecture in praise of astronomy, dated 9 January 1620, also salutes the achievements of the medieval Oxford mathematicians, including the Merton *calculatores*.

54. Ussher (note 29), p. 374 (3 October 1626). Bainbridge eventually bequeathed all his papers, including his work in Arabic and Persian, to Ussher, which is why they are in Trinity College, Dublin (MSS 382–86) and not in the Savilian Library.

55. Aubrey (note 15), p. 147.

56. Correspondence (incoming letters and outgoing drafts) with Snell, Erpenius, Golius, and Hortensius can be found in Trinity College, Dublin (MS 382). For the observations printed by Boulliau, see *Astronomia Philolaica*, Paris (1645), 467, probably from letters initially addressed to Hevelius, and see also note 15. For the astronomical expedition, see Tyacke (note 15), pp. 82–3.

57. Ward (note 15), pp. 273–4.

58. John Bainbridge (ed.), *Procli Sphaera Ptolemæi de Hypothesibus Planetarum Liber Singularis, Nunc Primum in Lucem Editus, Cui Accessit Ejusdem Ptolemæi Canon Regnorum*, London (1620), sig. ¶3r–¶4r. The same preface confirms Briggs's editorship of the Euclid volume and his intentions to complete the project ('fecit facietque'). Briggs's 1626 letter to Holstenius (MS D'Orville 52, fol. 134) shows that he considered himself a collaborator with Bainbridge on the projected edition of Ptolemy, on which they were both still working in that year. They hoped that the famous publishing firm of Froben in Basel would take on the project.

59. Ussher (note 29), p. 374 (3 October 1626).

60. As note 59.

61. Wood (note 15), Vol. III, pp. 306–7; Ward (note 15), pp. 129–35, bulked out by discussion of his ancestors through want of material. Aubrey does not mention him, and almost no learned correspondence survives. Wood unverifiably claimed that Turner 'wrote many admirable things, but being too curious and critical, he could never finish them according to his mind, and therefore cancell'd them'.

62. Niccolò Cabeo, *Philosophia Magnetica* (Ferrara, 1629).

63. William Gilbert, *De Magnete, Magneticisque Corporibus, et de Magno Magnete Tellure* (On the Magnet and Magnetic Bodies, and on that Great Magnet, the Earth) (London, 1600).

64. See Ward (note 15), pp. 133–4.

65. For the sum of evidence on Turner and some hints about the fate of his papers, see Feingold (note 15), pp. 65–6, 69–70, 156, 164.

66. Tyacke (note 15), pp. 89–90.

67. See Wood (note 15), Vol. III, pp. 324–8; Smith (note 14), 'Clarissimi ac Doctissimi Viri, Joannis. . . Vita'; Ward (note 15), pp. 135–53. His works were edited by Thomas Birch in three volumes in 1737, with a life prefaced to the first volume. For more recent publications, see Toomer (note 51), pp. 127–42, Zur Shalev, 'Measurer of all things: John Greaves (1602–1652), the Great Pyramid, and early modern metrology', *Journal of the History of Ideas* 63 (2002), 555–75, and 'The travel notebooks of John Greaves', in Alastair Hamilton *et al.* (ed.), *The Republic of Letters and the Levant*, Brill (2005), 77–102.

68. Ussher (note 29), p. 374 (3 October 1626).

69. See Ward (note 5), p. 274.

The author of this chapter is grateful to Mordechai Feingold and Daniel Fried for comments on an earlier draft, and to Lynsey Darby, Archivist of St John's College, Cambridge, for communicating images of the pertinent documents in note 21.

CHAPTER 2

Further reading

A comprehensive discussion of John Wallis's life and work is

Jason M. Rampelt's *Distinctions of Reason and Reasonable Distinctions: The Academic Life of John Wallis (1616–1703)*, Brill (2019). [*Life*]

There is also an up-to-date and reliable overview of Wallis's scientific work by Christoph J. Scriba in the *Dictionary of Scientific Biography*.

A translation of, and commentary on, Wallis's *Arithmetica Infinitorum* appears in Jacqueline A. Stedall, *The Arithmetic of Infinitesimals: John Wallis 1656*, Springer (2004).

Wallis's algebra is discussed in her book
A Discourse Concerning Algebra: English Algebra to 1685, Oxford University Press (2002). [*Algebra*]

The Correspondence of John Wallis (1616–1703) is being edited in eight volumes by Philip Beeley and Christoph J. Scriba for Oxford University Press (2003–). [*Correspondence*]

Wallis's collected works appear in the three volumes (and supplement) of his *Opera Mathematica*, published at the Sheldonian Theatre Press in Oxford from 1693 to 1699. [*Works*]

A full and well-documented account by Mordechai Feingold of the context of mathematical and scientific studies over this period at Oxford forms Chapter 6 of *The History of the University of Oxford*, Vol. IV: *17th-Century Oxford* (ed. Nicholas Tyacke), Clarendon Press (1997).

Notes and references

1. Letter from John Wallis to John French, August 1650, *Correspondence*, Vol. I, 30.

2. Christoph J. Scriba, 'The Autobiography of John Wallis, F.R.S.', *Notes and Records of the Royal Society* 25 (1970), 17–46, on p. 40.

3. This pamphlet was *A Serious and Faithfull Representation of the Judgements of the Ministers of the Gospell Within the Province of London. Contained in a Letter from them to the General and his Counsel of War, Jan. 18, 1649.*

4. See Scriba (note 2), 27.

5. John Wallis, *Institutio Logicae. Ad Communes Usus Accommodata* (at the Sheldonian Theatre, Oxford, 1687). The two Cambridge theses referred to are Thesis prima, 'Propositio singularis, in dispositione syllogistica, semper habet vim universalis', on pp. 219–34, and Thesis tertia, 'Quantitas non differt realiter a re quanta', on pp. 255–62; *Works*, Vol. III, ii, 219–34, 255–62.

6. One of Wallis's most significant contributions to the Parliamentary cause in advance of his election to the Savilian professorship was his role in revealing the Presbyterian Conspiracy of 1649–51. See Philip Beeley, 'Breaking the code: John Wallis and the politics of concealment', in *G. W. Leibniz und der Gelehrtenhabitus. Anonymität, Pseudonymität, Camouflage* (ed. W. Li and S. Noreik), Böhlau Verlag (2016), 49–81, on pp. 61–3.

7. See Jason Rampelt, 'The last word: John Wallis on the origin of the Royal Society', *History of Science* 46 (2008), 177–201, and Philip Beeley, 'Eine Geschichte zweier Städte: Der Streit um die wahren Ursprünge der Royal Society', *Acta Historica Leopoldina* 49 (2008), 135–62.

8. William Oughtred, *Clavis Mathematicae*, Thomas Harper (London, 1631). Oughtred's original title was *Arithmeticae in Numeris et Speciebus Institutio: Quae tum Logisticae, tum Analyticae, atque adeo totius Mathematicae quasi Clavis est* (Calculation in Numbers and Letters: Which was the Key first to Arithmetic, then to Analysis and now to the Whole of Mathematics).
   The third edition, to which Wallis contributed, was published in 1652, and the first English edition, *The Key of the Mathematics new Filed* (by Robert Wood of Merton College) appeared in 1647.

9. See the letter from John Wallis to John Smith, 28? November/[8? December] 1648, *Correspondence*, Vol. I, 9; John Wallis to William Brouncker, 5/[15] December 1656, *Correspondence*, Vol. I, 235–6; Christoph J. Scriba, *Studien zur Mathematik des John Wallis (1616–1703)*, Steiner Verlag (1966), 17–18. Cubic equations were first solved by Italian mathematicians in the 16th century and were discussed in Girolamo Cardan's *Ars Magna* (The Great Art) of 1545.

10. Letter from John Wallis to Johannes Hevelius, 3/[13] April 1649, *Correspondence*, Vol. I, 10–11.

11. John Wallis, *De Sectionibus Conicis Tractatus*, Leonard Lichfield for Thomas Robinson (Oxford, 1655), 1–2; *Works*, Vol. I, 293, 295. See Helena M. Pycior, *Symbols, Impossible Numbers, and Geometric Entanglements: British Algebra through the Commentaries on Newton's Universal Arithmetick*, Cambridge University Press (1997), 121–2.

12. See Paolo Mancosu, *Philosophy of Mathematics and Mathematical Practice in the Seventeenth Century*, Oxford University Press (1996), 121–4, and Philip Beeley, 'Infinity, infinitesimals, and the reform of Cavalieri: John Wallis and his critics', *Infinitesimal Differences: Controversies between Leibniz and his Contemporaries* (ed. U. Goldenbaum and D. Jesseph), De Gruyter (2008), 31–52.

13. John Wallis, *Artihmetica Infinitorum*, Leonard Lichfield (Oxford, 1656).

14. Wallis's lecture notes, *In Elementa Euclidis*, are in the Bodleian Library, MS Don. d. 45.

15. John Wallis, *Mathesis Universalis: sive, Opus Arithmeticum Integrum, tum Philologice, tum Mathematice Traditum*, Leonard Lichfield for Thomas Robinson (Oxford, 1657); *Works*, Vol. I.

16. See Scriba (note 2), 41.

17. John Wallis, *Mathesis Universalis*, sig. b4v and 202–8; *Works*, Vol. I, 13 and 123–6.

18. Henry Savile, *Praelectiones Tresdecim in Principium Elementorum Euclidis*, John Lichfield & Jacob Short (Oxford, 1621), 140. One version of the parallel postulate states that if L is any line, and if P is any point that does not lie on L, then there is a unique line that is parallel to L and passes through P.

19. John Wallis, *De Postulato Quinto et Definition Quinta lib. 6. Euclidis; Disceptatio Geometrica, Works*, Vol. II, 665–78; Vincenzo de Risi, *Leibniz on the Parallel Postulate and the Foundations of Geometry*, Birkhäuser (2016), 9–17.

20. John Wallis, *Oratio Inauguralis: in Auditorio Geometrico, Oxonii, Habita; Ultimo die Mensis Octobris, Anno Ærae Christinae 1649. Quum Publicam Professionem Auspicatus est*, Leonard Lichfield for Thomas Robinson (Oxford, 1657), sig. c3v; *Works*, Vol. I, 8.

21. See Philip Beeley and Christoph J. Scriba, 'Disputed glory: John Wallis and some questions of precedence in seventeenth-century mathematics', *Kosmos und Zahl: Beiträge zur Mathematik- und Astronomiegeschichte, zu Alexander von Humboldt und Leibniz* (ed. H. Hecht, R. Mikosch, *et al.*), Steiner Verlag (2008), 275–99, and Jacqueline Stedall, 'John Wallis and the French: his quarrels with Fermat, Pascal, Dulaurens, and Descartes', *Historia Mathematica* 39 (2012), 265–79. In *De Cycloide et Corporibus Inde Genitis*, Wallis presented his account of Pascal's prize questions on the cycloid, including his own solution.

22. John Wallis, *Elenchus Geometriae Hobbianae*, H. Hall for John Crooke (Oxford, 1655), and Thomas Hobbes, *Six Lessons to the Professors of the Mathematiques*,

one of Geometry, the other of Astronomy: In the Chaires set up by the Noble and Learned Sir Henrey Savile, in the University of Oxford, Andrew Crooke (London, 1656). See also Douglas M. Jesseph, *Squaring the Circle: The War between Hobbes and Wallis*, University of Chicago Press (1999), and Rampelt (*Life*), pp. 191–208.

23. Wallis's response to the dispute was to pen a tract entitled *A Defense of the Royal Society, and the Philosophical Transactions, Particularly Those of July, 1670. In Answer to the Cavils of Dr. William Holder*, T. Moore (London, 1678). On this dispute, see Mordechai Feingold, 'The origins of the Royal Society revisited', in *The Practice of Reform in Health, Medicine, and Science, 1500–2000* (ed. M. Pelling and S. Mandelbrote), Ashgate (2005), 167–83.

24. See Benjamin Wardhaugh, 'Marcus Meibom, mathematician: the *De proportionibus dialogus* (1655) and its responses', in *Marcus Meibom: Studies in the Life and Works of a Seventeenth-Century Polyhistor* (ed. M. Lundberg and J. Kreslins), University of Chicago Press (2020).

25. See Philip Beeley, 'A philosophical apprenticeship: Leibniz's correspondence with the Secretary of the Royal Society, Henry Oldenburg', *Leibniz and his Correspondents* (ed. P. Lodge), Cambridge University Press (2004), 47–73.

26. See, for example, the letters from Henry Oldenburg to John Wallis, 13/[23] October 1674, and from John Wallis to Henry Oldenburg, 15/[25] October 1674, *Correspondence*, Vol. IV, 428–9.

27. For Wallis's contributions to the Royal Society on natural philosophy see Adam Richter, 'On food and fossils: natural philosophy, mathematics, and biblical history in the works of John Wallis', *The Seventeenth Century* 35 (2020), 77–104, and Philip Beeley, 'Physical arguments and moral inducements: John Wallis on questions of antiquarianism and natural philosophy', *Notes and Records of the Royal Society* 72 (2018), 413–30.

28. See Philip Beeley, 'Early physics', *The Oxford Handbook of Leibniz* (ed. M. R. Antognazza), Oxford University Press (2018), 290–303.

29. See the letter from John Wallis to Henry Oldenburg, 7/[17] April 1671, *Correspondence*, Vol. III, 445.

30. See Stedall (*Algebra*) and Philip Beeley and Christoph J. Scriba, 'Wallis, Leibniz und der Fall von Harriot und Descartes: Zur Geschichte eines vermeintlichen Plagiats im 17. Jahrhundert', *Acta Historica Leopoldina* 45 (2005), 115–29.

31. John Wallis, *A Treatise of Algebra, both Historical and Practical. Shewing, the Original, Progress, and Advancement thereof, from time to time; and by what steps it hath attained to the heighth at which now it is*, John Playford for Richard Davis (London, 1685), 125–30, and *passim*.

32. Wallis's constructions for square roots appear in *A Treatise of Algebra* (note 31), pp. 266–7.

33. The full title is: John Wallis (ed.), *Jeremiæ Horroccii, Liverpoliensis Angli, ex Palatinatu Lancastriæ, Opera Posthuma: Viz. Astronomia Kepleriana, Defensa & Promota. Excerpta ex Epistolis ad Crabtræum Suum. Observationum Cœlestium Catalogus. Lunæ Theoria Nova. Accedunt Guilielmi Crabtræi, Mancestriensis, Observationes Cœlestes. Quibus Accesserunt, Johannis Flamstedii, Derbiensis, De Temporis Aequatione Diatriba. Numeri ad Lunæ Theoriam Horroccianam* (London, 1673).

34. See Michael D. Reeve, 'John Wallis, editor of Greek mathematical texts', in *Editing Texts/Texte edieren* (ed. G. W. Most), Vandenhoeck & Ruprecht (1998), 77–93.

35. See Philip Beeley, 'A designe Inchoate: Edward Bernard's planned edition of Euclid and its scholarly afterlife in late seventeenth-century Oxford', *Reading Mathematics*, 192–229.

36. See the collection of texts contained in *John Wallis: Writings on Music* (ed. D. Cram and B. Wardhaugh), Ashgate, 2014.

37. Charles Harding Firth, 'Thomas Scot's account of his actions as intelligencer during the Commonwealth', *English Historical Review* 12 (1897), 116–126, on p. 121.

38. See Philip Beeley, 'Un des mes amis: on Leibniz's relation to the English mathematician and theologian John Wallis', in *Leibniz and the English-speaking World* (ed. P. Phemister and S. Brown), Springer (2007), 63–81, and A. Ram 'Leibniz and the Royal Society revisited', in *Leibniz's Legacy and Impact* (ed. J. Weckend and L. Strickland), Routledge (2020), 23–52.

39. See Anthony Wood, 'On Wallis's Election to Custos Archivorum', February 1657/8, *Correspondence*, Vol. I, 396. Not only was the election itself questionable, there were complaints that holding the post of archivist was not compatible with the Savilian statutes. See John Wallis, 'Reasons shewing the consistency of the place of Custos Archivorum with that of a Savilian professor', February 1657/8, *Correspondence*, Vol. I, 394–5.

40. See Philip Beeley and Siegmund Probst, 'John Wallis (1616–1703): mathematician and divine', *Mathematics and the Divine* (ed. L. Bergmans and T. Koetsier), Elsevier (2005), 441–57.

41. See Scriba (note 2), 42–3.

CHAPTER 3

Further reading

Parts of this chapter are based heavily on the following chapters of *Oxford Figures*: 'Edmond Halley' and 'Oxford's Newtonian school' (by Allan Chapman), and 'Georgian Oxford' (by John Fauvel).

Brief biographical sketches of the Savilian professors in this chapter can be found in Joseph Foster (ed.), *Alumni Oxonienses, The Matriculation Registers of the University Arranged, Revised and Annotated*, 4 vols (Oxford, 1891).

For greater detail on Halley, Bliss, Robertson, and Rigaud, see the *Oxford Dictionary of National Biography* (2004), with further information on Halley in the *Dictionary of Scientific Biography*.

Three book-length biographies of Halley are:

Colin Ronan, *Edmond Halley: Genius in Eclipse*, Macdonald (1970) [*Ronan*]

Alan Cook, *Edmond Halley: Charting the Heavens and the Seas*, Clarendon Press (1997) [*Cook*] and Angus Armitage, *Edmond Halley*, Nelson (1966).

E. F. MacPike's *Correspondence and Papers of Edmond Halley*, Oxford University Press (1932), provides useful primary material.

For Nathaniel Bliss, see also the *Biographical Encyclopedia of Astronomers*, Springer (2007).

For John Smith, see James Beattie, *James Beattie's London Diary, 1773* (ed. R. S. Walker), Aberdeen University Press (1946), 131, and William Innes Addison, *The Snell Exhibitions: From the University of Glasgow to Balliol College, Oxford*, J. MacLehose & Sons (1901), 45.

For Abraham Robertson and Stephen Rigaud, see additionally 'Abram Robertson, D.D. [obituary]' and 'Professor Rigaud [obituary]', *The Gentleman's Magazine, and Historical Review* 97/1 (1827), 176–8, and (May 1839), 542–3, and also John Rigaud, *Stephen Peter Rigaud, M.A., F.R.S., Formerly Savilian Professor of Astronomy and Radcliffe Observer, Oxford: A Memoir*, privately printed, Oxford (1883).

For further background material, particularly on the astronomical context of 17th-century Oxford, see Lesley Murdin's *Under Newton's Shadow: Astronomical Practices in the Seventeenth Century*, Adam Hilger (1985), and the chapters by A. G. MacGregor and A. J. Turner on 'The Ashmolean Museum' and by G. L'E. Turner on 'The physical sciences' in *The History of the University of Oxford*, Vol. V: *The Eighteenth Century* (ed. L. S. Sutherland and L. G. Mitchell), Clarendon Press (1986). [*Sutherland & Mitchell*]

See also *Gunther*, Vol. 11, 1937.

Notes and references

1. Fauvel, 'Georgian Oxford', p. 185.

2. Quoted in *Ronan*, p. 6.

3. John Wallis (ed.), *Jeremiæ Horroccii, Liverpoliensis Angli, ex Palatinatu Lancastriæ, Opera Posthuma . . .* (London, 1673).

4. Translated from the original Latin by *Cook*, p. 53.

5. Flamsteed castigated Halley in a suppressed section of his *Historia Coelestis Britannica*, 3 vols (London, 1725), the original manuscript of which is in the Royal Society Library; see Allan Chapman (ed.), *The Preface to John Flamsteed's Historia Coelestis Britannica, or, British Catalogue of the Heavens (1725)*, National Maritime Museum Monograph 52 (1982), 160–80.

6. *Ronan*, p. 124; this quotation appears in Ronan without an exact source and comes from one of the three unpublished drafts of the same letter from Flamsteed to Newton, 20–22 February 1691, Royal Observatory Ms. 42, p. 129; for details, see H. W. Turnbull, *Correspondence of Sir Isaac Newton*, Vol. III, Cambridge University Press (1961), 199–205, although the version of the letter published (the final draft of 24 February 1691) omits this quotation.

7. Quoted by Stephen Jordan Rigaud (S. P. Rigaud's son) in *A Defence of Halley against the Charge of Religious Infidelity* (Oxford, 1844), 17.

8. Edmond Halley, 'Some remarks of the ancient state of the city of Palmyra', *Phil. Trans.* 19/218 (1695), 160–75.

9. Edmond Halley, 'An account of the several species of infinite quantity, and of the proportions they bear one to the other, as it was read before the Royal Society' and 'An estimate of the degrees of the mortality of mankind; drawn from curious tables of the births and funerals at the city of Breslaw; with an attempt to ascertain the price of annuities upon lives', *Phil. Trans.* 17/195 (1692), 556–8, and 17/196 (1693), 596–610. On the latter, see David R. Bellhouse, 'A new look at Halley's life table', *Journal of the Royal Statistical Society A* 174/3 (2011), 823–32.

10. Edmond Halley, 'An account of the causes of the late remarkable appearance of the planet Venus, seen this summer, for many days together, in the day time', *Phil. Trans.* 29/349 (1716), 466–8, on p. 466.

11. For more information on Halley as Savilian Professor, see *Cook*, Chapter 12.

12. H. Cook, 'The election of Edmond Halley to the Savilian Professorship of Geometry', *Journal for the History of Astronomy* 15 (1984), 34–6.

13. Letter from John Flamsteed to Abraham Sharp, 18 December 1703, in *The Correspondence of John Flamsteed, the First Astronomer Royal*, Vol. III, 1703–1719 (compiled and edited by Eric G. Forbes and for Marian Forbes by L. Murdin and F. Willmoth), Institute of Physics (Bristol), 2002, letter 922.
By this time, Halley had indeed been commissioned as a Royal Navy captain.

14. Letter of Thomas Hearne, 8 June 1704, in the Bodleian Library; cited in *Cook*, p. 322.

15. Edmond Halley, 'Astronomiæ cometicæ synopsis', *Phil. Trans.* 24/297 (1705), 1882–99.

16. *Apollonii Pergæi de Sectione Rationis Libri Duo ex Arab. MS. Lat. Versi. Accedunt Ejusdem de Sectione Spatii Libri Duo Restituti, Cum Lemmatibus Pappi ad Hos Apollonii Libros, Opera & Studio E. Halley* (Oxford, 1706).

17. Michael N. Fried, *Edmond Halley's Reconstruction of the Lost Book of Apollonius's Conics: Translation and Commentary*, Springer (2011).

18. *Apollonii Pergaei Conicorum Libro Octo, et Sereni Antissensis De Sectione Cylindri & Coni Libri Duo* (Oxford, 1710).

19. *Cook*, p. 340.

20. *Menelai Sphæricorum Libri III* (Oxford, 1758). On *Spherics*, see Roshdi Rashed and Athanase Papadopoulos, *Menelaus'* Spherics: *Early Translation and al-Mahani/al-Harawi's Version, Critical Edition, with Historical and Mathematical Commentaries*, De Gruyter (2017).

21. John Kersey, *The Elements of that Mathematical Art commonly called Algebra, expounded in two books, to which is added, Lectures read in the School of Geometry in Oxford, Concerning the Geometrical Construction of Algebraical Equations; And the Numerical Resolution of the same by the Compendium of Logarithms by Dr. Edmund Halley, Savilion Professor of Geometry in the University of Oxford* (London, 1717).

22. Edmond Halley, 'De constructionem Problematum solidorum, sive aequationum tertiae vel quartiae potetatis unica data parabola ac circulo efficienda', 'De numero radicum in æquationibus solidis ac biquadraticis, five tertiæ ac quartæ potestatis, carumque limitibus', and 'Methodus nova accurata & facilis inveniendi radices æqnationum quarumcumque generaliter, sine praviæ reductione', *Phil. Trans.* 16/179 (1687), 335–43 and 387–402, and 18/207 (1694), 136–48.

    An English translation of the last one is 'A new, exact, and easy method of finding the roots of any equations generally, and that without any previous reduction', C. Hutton, G. Shaw, and R. Pearson (eds), *The Philosophical Transactions of the Royal Society of London, From Their Commencement, in 1665, to the Year 1800: Abridged, With Notes and Biographical Illustrations*, Vol. III (London, 1809), 640–9.

    The 1694 paper was also printed as an appendix to Isaac Newton's *Arithmetica Universalis* (Cambridge, 1707).

23. Letter of Richard Waller to Owen Lloyd, 6 February 1694, quoted in *Cook*, p. 188.

24. See H. E. Bell, 'The Savilian Professors' houses and Halley's observatory at Oxford', *Notes and Records of the Royal Society* 16/2 (1961), 179–86, and *Gunther*, Vol. II, pp. 82–5.

25. Derek Howse, *Greenwich Observatory*, Vol. III, *The Buildings and Instruments*, Taylor & Francis (1975), 21, 32, 126.

26. *Gunther*, Vol. XI, p. 140.

27. Royal Society Archives: Certificates of election and candidature for Fellowship of the Royal Society, EC/1742/11.

28. 'Observations on the Transit of Venus over the Sun, on the 6th of June 1761: In a Letter to the Right Honourable George Earl of Macclesfield, President of the Royal Society, from the Reverend Nathaniel Bliss, M.A. Savilian Professor of Geometry in the University of Oxford, and F.R.S.', *Phil. Trans.* 52 (1761), 173–7.

29. Thomas Hornsby (ed.), *Astronomical Observations Made at the Royal Observatory at Greenwich From the Year MDCCL to the Year MDCCLXXII by the Rev. James Bradley*, Vol. II (Oxford, 1805), 345–420.

30. *Gunther*, Vol. XI, pp. 275–7. See also Allan Chapman, 'The King's Observatory at Greenwich and the First Astronomers Royal: Flamsteed to Bliss', in *Mathematics at the Meridian: The History of Mathematics at Greenwich* (ed. R. Flood, T. Mann, and M. Croarken), Chapman Hall/CRC Press (2019), 17–44, on pp. 41–3.

31. Letter from Jeremy Bentham to Jeremiah Bentham, 15 March 1763, *The Correspondence of Jeremy Bentham*, Vol. I: 1752–76 (ed. T. L. S. Sprigge), Athlone Press (1968), 67.

32. *Gunther*, Vol. XI, p. 277.

33. Joseph Foster (ed.) *Alumni Oxonienses*, Later Series A–D (1888), 104.

34. William Gardiner, *Tables of Logarithms, for All Numbers from 1 to 102100, and for the Sines and Tangents to Every Ten Seconds of Each Degree in the Quadrant; As also, for the Sines of the First 72 Minutes to Every Single Second: With Other Useful and Necessary Tables* (London, 1742).

35. 'A letter from the Rev. Mr. Joseph Betts, M. A. and Fellow of University College, Oxon, to Martin Folkes, Esq; Pr. R. S. containing observations on the late comet, made at Sherborn and Oxford; with the elements for computing its motions', *Phil. Trans.* 43/474 (1744), 91–100.

36. In his application for the position of Astronomer Royal, Betts received testimonials from the University of Oxford; see, for example, Derek Howse, 'Nevil Maskelyne, the Nautical Almanac, and G.M.T.', *Journal of Navigation* 38/2 (1985), 159–77.
His application for the Savilian Chair of Astronomy was supported by the Earl of Lichfield (Chancellor of Oxford University), the Earl of Bute (former Prime Minister), and the Earl of Halifax (Secretary of State); he thanked them in a dedication to a 1764 engraving, depicting the path of an annular solar eclipse of that year (History of Science Museum (Oxford), Radcliffe Observatory Archive, inventory number 27, 158).

37. R. J. P. Williams, Allan Chapman, and J. S. Rowlinson, *Chemistry at Oxford: a History from 1600 to 2005*, Royal Society of Chemistry (2009), 64.
A record of Smith's anatomy lectures survives in manuscript form: George Wingfield, *Anatomical Lectures Just as they were Taken at a Course read by Doctor Smith of St Mary's Hall, Oxford in the Laboratory There, 1759*, Bodleian Library,

MS Add A 302. See also *Gunther, Vol. I: Chemistry, Mathematics, Physics and Surveying*, pp. 58–9.

38. Alexander Carlyle, *Autobiography of the Rev. Dr. Alexander Carlyle, Minister of Inveresk: Containing Memorials of the Men and Events of his Time* (Edinburgh, 1860), 198.

39. Christopher Brooke and Roger Highfield, *Oxford and Cambridge*, Cambridge University Press (1988), 241.

40. See W. F. Sedgwick, rev. Alan Yoshioka, 'Robertson, Abram (1751–1826), mathematician and astronomer' *Oxford Dictionary of National Biography* (23 September 2004).
    The term 'sub-lecturer' appears to have been used in this context; see, for example, a letter from Lewis Evans to Charles Hutton of 23 May 1791, in which Robertson is referred to in this capacity: letter 53 in *The Correspondence of Charles Hutton (1737–1823): Mathematical Networks in Georgian Britain* (ed. B. Wardhaugh), Oxford University Press (2017), 81–2.

41. William Austin, *An Examination of the First Six Books of Euclid's Elements* (Oxford, 1781), i.

42. J. Smith, *Observations on the Use and Abuse of the Cheltenham Waters, in Which are Included Occasional Remarks on Different Saline Compositions* (Cheltenham, 1786).

43. See Bell (note 24), p. 184.

44. 'Dr. Gregory's Method for Teaching Mathematicks', in M. G. (ed.), *Mercurius Oxoniensis, or the Oxford Intelligencer* (London, 1707), 26–9.

45. John Caswell, *A Brief (but Full) Account of the Doctrine of Trigonometry, Both Plain and Spherical* (London, 1685); 'A letter of Dr Wallis to Dr Sloan, concerning the quadrature of the parts of the lunula of Hippocrates Chius, Performed by Mr John Perks; with the further improvements of the same, by Dr David Gregory, and Mr John Caswell', *Phil. Trans.* 21/259 (1699), 411–18.

46. See, for example, Keill's *Introductio ad Veram Physicam. Seu Lectiones Physicæ. Habitæ in Schola Naturalis Philosophiæ Academiæ Oxoniensis, Quibus Accedunt Christiani Hugenii Theoremata de Vi Centrifuga & Motu Circulari Demonstrata* (Oxford, 1702), later translated into English as *An Introduction to Natural Philosophy: or, Philosophical Lectures Read in the University of Oxford, Anno Dom. 1700: To Which are Added, the Demonstrations of Monsieur Huygens's Theorems, Concerning the Centrifugal Force and Circular Motion* (London, 1720).

47. Letter from Jonathan Swift to Robert Hunter, 12 January 1709, in *Correspondence of Jonathan Swift*, Vol. I (ed. H. Williams), Oxford University Press (1963), 121.

48. J. T. Desaguliers, Preface to *A Course of Experimental Philosophy* (London, 1734).

49. J. Keill, *Introductio ad Veram Astronomiam, Seu Lectiones Astronomicæ: Habitæ in Schola Astronomica Academiæ Oxoniensis* (Oxford, 1718); *Trigonometriæ Planæ & Sphæricæ Elementa. Item de Natura et Arithmetica Logarithmorum Tractatus Brevis* (Oxford, 1715).

50. 'An abstract of twenty lectures given by Dr Bradley at Oxford in 1747, transcribed by one of the auditors', Anonymous manuscript notes, Manuscript 3, History of Science Museum, University of Oxford.

51. Letter from George Carter to Archbishop Wake, 7 September 1721, Wake MS 16, f. 88, cited in *Sutherland & Mitchell*, p. 481.

52. Nicholas Amhurst, *Terrae Filius* (1721), no. 21.

53. The latter had appeared first in Latin in 1715 as *Euclidis Elementorum, Libri Priores Sex, Item Undecimus & Duodecimus*, before being translated into English in 1723 with an English version of the trigonometrical text appended: *Euclid's Elements of Geometry from the Latin Translation of Commandine. To Which is Added, a Treatise of the Nature and Arithmetic of Logarithms; Likewise Another of the Elements of Plain and Spherical Trigonometry.*

54. Letter from Charles James Fox to Sir George Macartney, 13 February 1765, *Memorials and Correspondence of Charles James Fox*, Vol. I (ed. Lord John Russell) (London, 1853), 19.

55. As note 54.

56. See Russell (note 54), p. 20. On the other hand, Newcome's remark about delaying geometrical studies until Fox's return may have been sarcastic.

57. Love issued a robust response in a pamphlet of his own entitled *I am at a Loss*, Bodleian Library, Gough Oxf. 90 (10).

58. P. Quarrie, 'The Christ Church collections books', *Sutherland & Mitchell*, 511.

59. See Keith Hannabuss, 'Mathematics in Victorian Oxford', *Mathematics in Victorian Britain* (ed. R. Flood, A. Rice, and R. Wilson), Oxford University Press (2011), 35–50.

60. Abram Robertson, *A Demonstration of the Fifth Definition of the Fifth Book of Euclid*, Oxford, 1789.

61. Abram Robertson, *Sectionum Conicarum Libri Septem: Accedit Tractatus de Sectionibus Conicis, et de Scriptoribus qui Earum Doctrinam Tradiderunt*, and *Archimedis Quæ Supersunt Omnia Cum Eutocii Ascalonitæ Commentariis. Ex Recensione Josephi Torelli, Veronensis, Cum Nova Versione Latina. Accedunt Lectiones Variantes Ex Codd. Mediceo Et Parisiensibus* (Oxford, 1792).

62. Royal Society Archives: Certificates of election and candidature for Fellowship of the Royal Society, EC/1795/14.

63. Abram Robertson, *A Geometrical Treatise of Conic Sections: In four books. To Which is Added a Treatise on the Primary Properties of Conchoids, the Cissoid, the Quadratrix, Cycloids, the Logarithmic Curve, and the Logarithmic, Archimedean, and Hyperbolic Spirals* (Oxford, 1802), and *Elements of Conic Sections Deduced from the Cone: and Designed as an Introduction to the Newtonian Philosophy* (Oxford, 1818).

64. 'The binomial theorem demonstrated by the principles of multiplication', and 'A new demonstration of the binomial theorem, when the exponent is a positive or negative fraction', *Phil. Trans.* 85 (1795), 298–321, and 96 (1806), 305–26.

65. See R. W., 'Philosophical Transactions of the Royal Society of London, for the Year 1806', *Monthly Review, or Literary Journal* 52 (February 1807), 159–72, 163–5, Abram Robertson, *A Reply to a Critical and Monthly Reviewer, in Which is Inserted Euler's Demonstration of the Binomial Theorem* (Oxford, 1808), and R. W., 'A Reply to a Monthly Reviewer . . . [review]', *Monthly Review, or Literary Journal* 56 (June 1808), 134–50.

66. 'Robertson [obituary]', *Gentleman's Magazine*, p. 176.

67. Quoted in Fauvel, 'Georgian Oxford', p. 196.

68. 'On the precession of the equinoxes', *Phil. Trans.* 97 (1807), 57–82.

69. Robertson's reputation as an observer has become tainted in recent decades by the revelation that he fabricated some observational data; see E. Myles Standish, 'Fabricated transit data by Abram Robertson', *DIO* 7/1 (1997), 3–13.

70. See, for example, 'Direct and expeditious methods of calculating the excentric from the mean anomaly of a planet', and 'Demonstrations of the late Dr. Maskelyne's formulae for finding the longitude and latitude of a celestial object from its right ascension and declination; and for finding its right ascension and declination from its longitude and latitude, the obliquity of the ecliptic being given in both cases', *Phil. Trans.* 106 (1816), 127–37 and 138–48.

71. Quoted in 'Rigaud [obituary]', *Gentleman's Magazine*, p. 543.

72. Royal Society Archives: Certificates of election and candidature for Fellowship of the Royal Society, EC/1805/04.

73. As note 71.

74. Stephen Peter Rigaud, *Syllabus of a Course of Lectures on Experimental Philosophy* (Oxford, 1821).

75. Roger Hutchins, 'Rigaud, Stephen Peter ('1774–1839), astronomer', *Oxford Dictionary of National Biography* (23 September 2004).

76. Allan Chapman, 'Thomas Hornsby and the Radcliffe Observatory', *Oxford Figures*, 203–20, on p. 218.

77. A near-contemporary account of the non-publication of the papers may be found in a letter written by Rigaud to the editors of the *Journal of the Royal Institution of Great Britain* 2 (1831), 267–71.

78. Fauvel, 'Georgian Oxford', p. 192.

79. Stephen Peter Rigaud, 'On Harriot's astronomical observations contained in his unpublished manuscripts belonging to the Earl of Egremont [abstract, 17/24 May 1832]', *Proceedings of the Royal Society of London* 3 (1837), 125–6; *Report of the First and Second Meetings of the British Association for the Advancement of Science; at York in 1831, and at Oxford in 1832: including its Proceedings, Recommendations, and Transactions* (London, 1833), 602.

80. The full publication of Harriot's papers, under the editorship of Matthias Schemmel, Jacqueline Stedall, and Robert Goulding, has become possible in recent years, thanks to the flexibility afforded by an online format: *https://echo.mpiwgberlin.mpg.de/content/scientific_revolution/harriot*

81. Bodleian Library, Mss. Rigaud 45 and 53; see also S. P. Rigaud, 'Some account of those manuscripts in Great Britain, which contain the Greek text of the *Mathematical Collections* of Pappus', *Edinburgh Philosophical Journal* 7 (1822), 56–64, 219–25.

82. Stephen Peter Rigaud (ed.), *Historical Essay on the First Publication of Sir Isaac Newton's Principia, Miscellaneous Works and Correspondence of the Rev. James Bradley*, and *Supplement to Dr. Bradley's Miscellaneous Works: with an Account of Harriot's Astronomical Papers* (Oxford, 1838, 1832, and 1833).

83. Stephen Peter Rigaud, *Correspondence of Scientific Men of the Seventeenth Century: Including Letters of Barrow, Flamsteed, Wallis, and Newton, printed from the Originals in the Collection of the Right Honourable the Earl of Macclesfield* (Oxford, 1841).

84. Stephen Peter Rigaud, 'Some particulars respecting the principal instruments at Greenwich in the time of Dr. Halley', *Monthly Notices of the Royal Astronomical Society* 3/23 (May 1836), 193–4.

85. Stephen Peter Rigaud, 'Some particulars of the Life of Dr. Halley', *Monthly Notices of the Royal Astronomical Society* 3/10 (December 1834), 67–8, and *Some Account of Halley's* Astronomiæ Cometicæ Synopsis, *which contains his Investigations of the Orbits of Comets* (Oxford, 1835).

86. Quoted in Rigaud [note 75]. Rigaud's scholarship on Halley was continued by his son (see note 7).

The authors of this chapter are grateful to Michael Riordan, Archivist of The Queen's College for supplying the signature of Halley.

CHAPTER 4

Further reading

Much useful background information on Oxford University in the 19th century can be found in:

W. R. Ward, *Victorian Oxford*, Frank Cass (1965) [*Ward*]

A. J. Engel, *From Clergyman to Don*, Oxford University Press (1983) [*Engel*]

and, from a more personal viewpoint,

Revd W. *Tuckwell's Reminiscences of Oxford*, Cassell (1901) [*Tuckwell*]

(William Tuckwell was, for a while, Headmaster of New College School, Oxford.)

A good biography of Baden Powell, more slanted towards theology than mathematics, is

P. Corsi, *Religion and Science*, Cambridge University Press (1988).

There is also useful information in W. Tuckwell's *Pre-Tractarian Oxford*, Cassell (1909).

There is no full-length biography of Henry Smith, but biographical notes are included in

The Collected Mathematical Papers of Henry John Stephen Smith M.A., F.R.S. (ed. J. W. L. Glaisher), 2 vols (Oxford, 1894), reprinted by Chelsea in 1965. [*Smith papers*]

This collection includes Smith's *Report on the Theory of Numbers*, and biographical sketches by his friends and colleagues Charles Pearson, Benjamin Jowett, and James Glaisher; the story of the Paris Grand Prix is described in Glaisher's contribution. There is also a chapter on Henry Smith by Keith Hannabuss in *Oxford Figures*.
For Nevil Story Maskelyne and his circle, and useful information about Smith's friends and acquaintances, see Vanda Morton, *Oxford Rebels*, Alan Sutton (1987) [*Morton*]

Notes and references

1. See Tanis Hinchcliffe, *North Oxford*, Yale University Press (1992).

2. *Tuckwell*, p. 4.

3. See *Engel*, pp. 34, 56, and *Ward*, p. 87.

4. See *Ward*, pp. 260ff.

5. Baden Powell, *On the Present State and Future Prospects of Mathematical and Physical Studies at the University of Oxford* (1832), 27.

6. See Powell (note 5), p. 40.

7. James Pycroft, *Oxford Memories* (London, 1886).

8. The British Association for the Advancement of Science was renamed the British Science Association (BSA) in 2009.

9. J. Morrell and A. Thackray, *Gentlemen of Science. Early Years of the British Association for the Advancement of Science*, Clarendon Press (1981), 386–96.

10. Baden Powell, 'Report on the present state of our knowledge of radiant heat', *Proceedings of the British Association for the Advancement of Science Meeting* (1832), 259–301.

11. It was in Baden Powell's house, during the 1847 meeting, that John Couch Adams and Urbain Le Verrier met for the first time.

12. As a young student at Bristol College, George Gabriel Stokes had been less impressed by a couple of high Cambridge Wranglers than by the Oxford graduate Francis Newman. Newman, the younger brother of John Henry Newman, had been an exceptional student at Oxford, graduating in 1826 just before Powell returned as Savilian Professor. Perhaps Oxford mathematics was in less bleak a state than Powell had suggested.

13. E. G. W. Bill, *University Reform in Nineteenth-Century Oxford*, Clarendon Press (1973), 90.

14. For the status of the so-called Wilberforce–Huxley debate, see Allan Chapman's observations in 'Monkeying about with History: Remembering the 'Great Debate' of 1860', and 'Aping Our Ancestors': how the 'Great Debate' of 1860 was invented' in the Oxford Magazine for Trinity Term, 2010: Noughth Week, pp. 10–12, and Eighth Week, pp. 17–18.

15. Baden Powell, 'On the study of the evidences of Christianity', *Essays and Reviews* (1860), 94–144, on p. 139.

16. Charles Darwin's *On the Origins of Species* was published in 1859, and the third edition (which included the tribute to Baden Powell) appeared in 1861.

17. This negative view of Baden Powell's article appears in Geoffrey Faber, *Jowett: A Portrait with Background*, Faber & Faber (1957).

18. G. C. Smith, *The Boole–De Morgan Correspondence 1842–1864*, Oxford University Press (1982), 83.

19. D. MacHale, *George Boole*, Boole Press (1985), 168.

20. Letter to Eleanor Smith, *Biographical Sketches and Recollections (with early letters) of Henry John Stephen Smith M.A., F.R.S.*, printed for private circulation (1894), 38–9.

21. See T. Smith, 'The Balliol–Trinity Science laboratories', in *Balliol Studies* (ed. J. M. Prest), Leopard's Head Press (1982), 190.

22. Morton, pp. 55 and 95.

23. R. J. P. Williams, J. S. Rowlinson, and A. Chapman, *Chemistry at Oxford: A History from 1600 to 2005*, Royal Society of Chemistry (2009), 118–27.

24. See David B. *Wilson, Kelvin and Stokes: A Comparative Study in Victorian Physics*, Adam Hilger (1987), and M. McCartney, A. Whitaker, and A. Wood (eds), *George Gabriel Stokes: Life, Science and Faith*, Oxford University Press (2019).

25. John Wellesley Russell, *An Elementary Treatise on Pure Geometry: With Numerous Examples* (Oxford, 1893).

26. Evelyn Abbott and Lewis Campbell, *The Life and Letters of Benjamin Jowett, M.A.*, Vol. II, E. P. Dutton and Co. (1997), 238.

27. Smith papers, Vol 1, p. xix.

28. G. Kowalewski, *Bestand und Wandel*, Oldenbourg Verlag (1949), 187.

29. Eleanor Smith, letter to Charles Pearson, 29 October 1874, Bodleian Library, Pearson Collection, Eng. lett. d. l91.

30. A number is transcendental if it is not a root of any polynomial equation with integer coefficients.

31. N. Kurti, 'Opportunity lost in 1865?', *Nature* 308 (1984), 313–14.

32. Charles L. Dodgson's *An Elementary Treatise on Determinants*, published by Macmillan in 1867, was the book that he was rumoured to have sent to Queen Victoria, following her request for his next publications after *Alice's Adventures in Wonderland*.

33. Tony Crilly, *Arthur Cayley: Mathematical Laureate of the Victorian Age*, Johns Hopkins University Press (2006), 230, 281, and 520.

34. Maryna Viazovska, 'The sphere-packing problem in dimension 8', *Annals of Mathematics* 185 (2017), 991–1015.

35. Smith papers, Vol 1, p. xxxxiii.

36. Radu Coldea *et al.*, 'Quantum criticality in an Ising chain: experimental evidence for emergent E8 symmetry', *Science* 327 (2010), 177–80.

37. H. Colquhoun, R. Grau-Crespo, *et al.*, 'Elements of fractal geometry in the 1H NMR spectrum of a copolymer intercalation-complex: identification of the underlying Cantor set', and 'Single-site binding of pyrene to poly(ester-imide)s incorporating long spacer-units; prediction of NMR resonance patterns from a fractal model', *Chemical Science* 9 (2018), 4052, and 11 (2020), 12165–77.

38. Smith papers, Vol 1, p. lxvi.

39. Smith papers, Vol 1, p. lxvii.

40. V. H. H. Green, *Oxford Common Room: A Study of Lincoln College and Mark Pattison*, Edward Arnold (1957), 306.

41. Smith papers, Vol 1, p. lxx.

42. Henry Smith, 'On the present state and prospects of some branches of mathematics', *Proceedings of the London Mathematical Society* 8 (1876), 6–29.

The author of this chapter wishes to thank Howard Colquhoun for drawing attention to his work.

CHAPTER 5

Further reading

Numerous biographical articles of Sylvester have appeared in the literature, but the only full-length biography is

Karen Hunger Parshall, *James Joseph Sylvester: Jewish Mathematician in a Victorian World*, Johns Hopkins University Press (2006). [*Parshall*]

Extracts from this book appear here with the kind permission of the publishers.

For a sampling of correspondence to and from Sylvester, with a detailed historical and mathematical commentary, see

Karen Hunger Parshall, *James Joseph Sylvester: Life and Work in Letters*, Clarendon Press (1998).[*Letters*]

Sylvester's mathematical works were published as

James Joseph Sylvester, *The Collected Mathematical Papers of James Joseph Sylvester*, 4 vols, Cambridge University Press (1904–12); reprint edn, Chelsea Publishing Company (1973). [*Works*]

Many of Sylvester's papers can be found in the library of St John's College, Cambridge. [*Papers*]

Notes and references

1. For more details about Sylvester's early years, see *Parshall*, pp. 16–25.

2. Percy MacMahon, 'James Joseph Sylvester', *Proceedings of the London Mathematical Society* 63 (1898), ix–xxv, on p. ix.

3. For the quotations see H. Hale Bellot, *University College London 1826–1926*, University of London Press (1929), 180, and a letter dated 25 February 1829 from Elias Sylvester (James's elder brother) to the University of London authorities, in College Correspondence, Sylvester, E. J. 1829: 1614 and 1615, University College London Archives.

4. Letter from William Leece Drinkwater to A. Theodore Brown, 24 April 1897, in A. Theodore Brown, *Some Account of the Royal Institution School of Liverpool with a Roll of Masters and Boys (1819 to 1892, A.D.)*, 2nd edn, University of Liverpool and Hodder and Stoughton (1927), 50–1.

5. For more details on Sylvester's time at Cambridge, see *Parshall*, pp. 26–48. Sylvester was out of residence at Cambridge for the two calendar years of 1834

and 1835, ostensibly due to illness, but probably also due to his father's death in 1834—this explains his six-year undergraduate association with the University.

6. For Sylvester's years on the faculty at University College, London, see *Parshall*, pp. 49–64.

7. For more on Sylvester's association with the University of Virginia and the United States between 1842 and 1844, see *Parshall*, pp. 64–80.

8. Letter from J. J. Sylvester to Joseph Henry, 12 April 1846, Record Unit 7001, Joseph Henry Collection, Smithsonian Institution Archives; in *Letters*, p. 15.

9. For more on Sylvester's return to London, see *Parshall*, pp. 81–106.

10. For Cayley's life and career, see Tony Crilly, *Arthur Cayley: Mathematician Laureate of the Victorian Age*, Johns Hopkins University Press (2006).

11. For a concise technical account of these developments, see Karen Hunger Parshall, 'The British development of the theory of invariants (1841–1895)', *BSHM Bulletin, British Society for the History of Mathematics* 21/3 (2006), 186–99.

12. Sylvester's years at Woolwich are treated in greater detail in *Parshall*, pp. 144–215.

13. Letter from J. J. Sylvester to Lord Brougham, 16 September 1855, in Sylvester, J. J., 20241, Brougham Papers, University College London Archives; in *Letters*, p. 90.

14. These 'uneasy years' in Sylvester's life are covered in *Parshall*, pp. 192–224.

15. Letter from J. J. Sylvester to Arthur Cayley, 19 February 1875, in Miscellaneous Manuscripts, MM.15.20, Royal Society of London.

16. Sylvester's years at the Johns Hopkins University are detailed in *Parshall*, pp. 226–73.

17. Letter from Arthur Cayley to J. J. Sylvester, 12 February 1883, in Papers, Box 2. What follows is adapted from *Parshall*, pp. 273–303.

18. Letter from J. J. Sylvester to Arthur Cayley, 16 March 1883, in *Papers*, Box 11, and *Letters*, p. 222. The quotations that follow in this paragraph are all from this letter.

19. Letter from J. J. Sylvester to Arthur Cayley, 10 April 1883, in Papers, Box 11, and Letters, pp. 225–7.

20. Letter from J. J. Sylvester to Arthur Cayley, 22 September 1883, in Papers, Box 11.

21. Cecil Roth, 'The Jews in English universities', *Miscellanies of the Jewish Historical Society of England*, Part 4, Jewish Historical Society of England (1942), 102–15, on p. 114.

22. Arthur J. Engle, 'Emerging concepts of the academic profession at Oxford 1800–1854', *The University in Society* (ed. Lawrence Stone), 2 vols, *Oxford and*

Cambridge from the 14th to the Early 19th Century, Vol. 1, Princeton University Press (1974), 305–51, on pp. 348–9.

23. Letter from J. J. Sylvester to Charles Taylor, 20 January 1884, in *Papers*, W.1.

24. Letter from J. J. Sylvester to Arthur Cayley, 29 January 1884, in *Papers*, Box 12, and *Letters*, pp. 242–3.

25. Letter from J. J. Sylvester to Arthur Cayley, 20 May 1884, in *Papers*, Box 12, and *Letters*, pp. 250–3.

26. As note 25.

27. Letter from J. J. Sylvester, 12 July 1884, in *Papers*, Box 12.

28. Letter from J. J. Sylvester to Arthur Cayley, 2 November 1884, in *Papers*, Box 12, and *Letters*, pp. 253–5. The text was Wilhelm Fiedler's German translation and adaptation of George Salmon's *A Treatise on Conic Sections*, originally published as *Analytische Geometrie der Kegelschnitte von George Salmon* (Leipzig, 1860).

29. As note 28; the quotations that follow in the next paragraph are also from this letter.

30. Pamphlet of four poems addressed 'To the Warden of New College with the author's kind regards', in the Archives of New College Library, Oxford.

31. Letter from J. J. Sylvester to Arthur Cayley, 22 April 1885, in *Papers*, Box 12.

32. As note 31.

33. It is not clear whether Cayley could make the trip. Sylvester had definitely invited him and was frustrated on 28 November when he wrote that 'you do not say . . . whether there is *any chance* of my having the pleasure of seeing you here for the lecture'. Letter from J. J. Sylvester to Arthur Cayley, 28 November 1885, in *Papers*, Box 12.

34. Sylvester, 'Inaugural Lecture at Oxford', in *Works*, Vol. 4, p. 280.

35. As note 34, p. 281; the next quotation is also from this page.

36. As note 34, p. 284; the quotations that follow in this paragraph are also on this page.

37. As note 34, p. 298; the next quotation is also from this page.

38. As note 34, p. 297.

39. As note 34, p. 300.

40. Letter from J. J. Sylvester to Arthur Cayley, 18 February 1886, in *Papers*, Box 12.

41. J. J. Sylvester, 'Lectures on the theory of reciprocants', *American Journal of Mathematics* 8 (1886), 196–260, 9 (1887), 1–37, 113–61, and 297–352, and 10 (1888), 1–16, or *Works*, Vol. 4, pp. 303–513.

42. As note 41, p. 303; for the references to the various papers, see *Letters*, p. 265.

43. Letter from Ludwig Brill to J. J. Sylvester, 26 November 1886, *Papers*, Box 2. For more information on the Brill models, see Gerd Fischer (ed.), *Mathematische Modelle*, 2 vols, Akademie Verlag (1986).

44. Letter from J. J. Sylvester to Daniel Coit Gilman, 11 March 1887, in Daniel Coit Gilman Papers, Coll #1 Corresp., Special Collections, Milton S. Eisenhower Library, The Johns Hopkins University, and *Letters*, p. 263; the quotes that follow in this paragraph are from this letter.

45. See the seven short papers in *Works*, Vol. 4, pp. 592–629.

46. See Edwin Bailey Elliott, 'Why and how the Society began and kept going', lecture at the 200th meeting of the Oxford Mathematical Society, 16 May 1925, privately printed pamphlet, 16 pp., on p. 12. The account of the formation and early years of the Society that appears here comes from this source.

47. As note 46, pp. 12–13.

48. Minutes, New College, Oxford, item 3507, p. 53, Archives, Library of New College, Oxford.

CHAPTER 6

Further reading

Hardy's mathematical writings were published in seven volumes, entitled

Collected Papers of G. H. Hardy, Clarendon Press (1966–79). [*Papers*]

His classic book, *A Mathematician's Apology*, was published by Cambridge University Press in 1940, and a later edition, with a Foreword by C. P. Snow, appeared in 1967. [*Apology*]. The page numbers below refer to the 1967 edition.

A great deal of information about the life and works of Hardy and his contemporaries can be found in

D. J. Albers, G. L. Alexanderson, and W. Dunham (eds), *The G. H. Hardy Reader*, Mathematical Association of America (2015). [*Reader*] See also

Adrian C. Rice and Robin J. Wilson, 'The rise of British Analysis in the early 20th century: the role of G. H. Hardy and the London Mathematical Society', *Historia Mathematica* 30 (2003), 173–94.

Hardy's collaboration with Littlewood is described in

B. Bollobas (ed.), *Littlewood's Miscellany*, Cambridge University Press (1986) [*Littlewood*], and in

Robin Wilson, 'Hardy and Littlewood', in *Cambridge Scientific Minds* (ed. P. Harman and S. Mitton), Cambridge University Press (2002), 202–19.

Information about Ramanujan and his relationship with Hardy appears in

R. Kanigel, *The Man who knew Infinity: A Life of the Genius Ramanujan*, Scribner's (1991). [*Kanigel*]

For some background on Oxford mathematics and mathematicians at this time, see Margaret E. Rayner's chapter, 'The 20th century', in *Oxford Figures*, and Chapter 8 of

Jack Morrell, *Science at Oxford 1914–1939: Transforming an Arts University*, Clarendon Press (1997). [*Morrell*]

Finally, Edward Titchmarsh wrote an extensive summary of his life; these were later extended by his wife. [*Notes*]

Notes and references

1. Keith Hannabuss, 'Mathematics in Victorian Oxford', *Mathematics in Victorian Britain* (ed. R. Flood, A. Rice, and R. Wilson), Oxford University Press (2011), 35–50, on p. 50.

2. *Apology*, pp. 148, 40.

3. At the end of 1919 Trinity College offered Russell a five-year lectureship in Logic and the Principles of Mathematics, which he accepted but never took it up, spending the academic year 1920–21 lecturing in China and Japan. Years later, Hardy recalled all these events in a short pamphlet, *Bertrand Russell and Trinity*, Cambridge University Press (1942); see also I. Grattan-Guinness, 'Russell and G. H. Hardy: a study of their relationship', *Russell, Journal of the Bertrand Russell Archives*, McMaster University Library Press (Winter 1991–92), 165–79.

4. *Littlewood*, p. 89.

5. Grattan-Guinness (note 3), p. 178.

6. At the time of Hardy's election in 1919, Micaiah J. M. Hill FRS was the first Astor Professor of Mathematics at University College, London, Percy MacMahon FRS was a combinatorialist who had been President of the London Mathematical Society, and Charles H. Sampson was mathematics tutor at Brasenose College and became the College's Principal in 1920.

7. See Grattan-Guinness (note 3), p. 175; further information can be found in the same author's article, 'A mathematical union: William Henry and Grace Chisholm Young', *Annals of Science* 29 (1972), 105–86, on pp. 161–2.

8. Hardy's inaugural address, 'Some famous problems of the Theory of Numbers, and in particular Waring's Problem. An Inaugural Lecture delivered before the University of Oxford', was printed by the Clarendon Press (1920) = *Papers*, Vol. I (1966), 647–78.

9. Waring's problem was answered in the affirmative by David Hilbert in 1909. For more information see W. J. Ellison, 'Waring's Problem', *American Mathematical Monthly* 78 (January 1971), 10–36.

10. The *Partitio Numerorum* papers appear in *Papers* Vol. I (1966), 382–530. Hardy and Ramanujan's paper was 'Asymptotic formulae in combinatory analysis', *Proceedings of the London Mathematical Society* (2) 17 (1918), 75–115 = *Papers*, Vol. I, pp. 306–39. It is discussed in Adrian Rice, 'Partnership, partition, and proof: The path to the Hardy–Ramanujan partition formula', *American Mathematical Monthly* 124 (January 2018), 3–15.

11. *Littlewood*, p. 118.

12. E. C. Titchmarsh, 'Godfrey Harold Hardy 1877–1947', *Obituary Notices of Fellows of The Royal Society* 6 (1949), 447–58, on p. 450 = *Papers*, Vol. I (1996), 1–12.

13. *New College, Oxford, 1379–1979* (ed. J. Buxton and P. Williams), published by the Warden and Fellows of New College (1979), 116–17.

14. Among Oxford's Presidents of the London Mathematical Society were E. B. Elliott (1896–98), A. E. H. Love (1912–14), A. L. Dixon (1924–26), and G. H. Hardy (1926–28); Dixon's brother, A. C. Dixon was also President (1931–33). After the presidency of E. A. Milne (1937–39), Hardy served a second term (1939–41), followed by J. E. Littlewood (1941–43) and by E. C. Titchmarsh (1945–47). The De Morgan Medal is awarded every third year in memory of Augustus De Morgan, the Society's founder and first President.

15. *Apology*, p. 147.

16. Letter from W. L. Ferrar to his Uncle Fred, 1913.

17. Verbal recollection by E. C. Thompson.

18. William Hunter McCrea, 'Edward Arthur Milne, 1896–1950', *Biographical Memoirs of Fellows of the Royal Society* 7/20 (November 1951), 420–43, on p. 428.

19. E. B. Elliott, 'Why and how the Society began and kept going', lecture to the Oxford and Mathematical and Physical Society (200th meeting) (16 May 1925), 8–9.

20. Titchmarsh (note 12), on p. 451.

21. See G. L. Alexanderson and D. H. Mugler, *Lion Hunting and Other Mathematical Pursuits*, Mathematical Association of America (1995), 235 = *Reader*, pp. 143–4.

22. Titchmarsh (note 12), on p. 451.

23. See James Tattersall and Shawnee McMurran, 'An interview with Dame Mary L. Cartwright, D.B.E., F.R.S.', *College Mathematics Journal* 32/4 (2001), 242–54.

24. These details about Hardy and Littlewood's collaboration can be found in *Miscellany*, pp. 2, 8–9.

25. Harald Bohr, *Collected Works*, Dansk Mat. Forening, Vol. 1 (1953), xxvii–xxviii = *Reader*, p. 9 and *Littlewood*, pp. 8–9.

26. See *Littlewood*, pp. 8–9.

27. See *Apology*, pp. 12, 29.

28. From 1934 Oxford's mathematicians occupied six rooms in the enlarged Radcliffe Science Library. By 1953 the department had increased in size and moved to 10 Parks Road. The first purpose-built Mathematical Institute in Oxford, at 24–29 St Giles, was opened in 1966, and the Institute's current home, the Andrew Wiles Building, is in the Radcliffe Observatory Quarter, Woodstock Road.

29. *Apology*, p. 149.

30. Dyson's letter to C. P. Snow, 22 May 1967, appears in *Kanigel*, p. 368, and *Reader*, p. 80.

31. *Morrell*, p. 312.

32. Letter from Hardy to F. Dewsbury, reprinted in B. C. Berndt and R. A. Rankin, *Ramanujan: Letters and Commentary, History of Mathematics Series 9, American and London Mathematical Societies* (1991), 243 = *Reader*, p. 75.

33. G. H. Hardy, 'Srinivasa Ramanujan', *Proceedings of the London Mathematical Society* (2) 19 (1921), xl–lviii = *Papers*, Vol. VII (1979), 702–20; the *Manchester Guardian* comment appeared on 2 December 1947.

34. P. V. Seshu Aiyah knew Ramanujan as a student in Ramanujan's home town of Kumbakonam in India, and later became a mathematics professor at the Presidency College in Madras (now Chennai). Bertram Martin Wilson FRSE taught mathematics at the University of Liverpool from 1920 to 1933, before being appointed professor at University College, Dundee; he died from an infection at the age of 38 while co-editing Ramanujan's notebooks.

35. G. H. Hardy, P. V. Seshu Aiyar, and B. M. Wilson (eds), *Collected Papers of Srinivasa Ramanujan*, Cambridge University Press (1927). See also *Kanigel*, p. 372, and *Reader*, p. 83.

36. Letter of 7 January 1919; see J. W. Dauben, 'Mathematicians and World War I: The international diplomacy of G. H. Hardy and Gösta Mittag-Leffler as reflected in their personal correspondence', *Historia Mathematica* 7 (1980), 261–88, on p. 264.

37. This remark has been quoted in several places—for example, *Kanigel*, p. 363, and *Reader*, p. 156.

38. Letter of 30 September 1921; see Dauben (note 36), on pp. 276–7. Further information about these Congresses and the political issues surrounding them can be found in Donald J. Albers, G. L. Alexanderson, and Constance Reid, *International Mathematical Congresses: An Illustrated History 1893–1986*, Springer-Verlag (1987), 16–21.

39. For more information about the Rockefeller Fellowships, see Reinhard Siegmund-Schultze, *Rockefeller and the Internationalization of Mathematics Between the Two World Wars*, Springer (2001).

40. G. H. Hardy, J. E. Littlewood, and G. Pólya, *Inequalities*, Cambridge University Press (1934).

41. G. H. Hardy, 'The theory of numbers', *British Association Report* 90 (1922), 16–24 = *Nature* 110 (16 September 1922), 381–5, and *Papers*, Vol. VII (1979), 514–18.

42. This photograph first appeared in the *Daily Mirror*, 11 August 1926.

43. This paraphrase is credited to J. B. S. Haldane in *Kanigel*, p. 366.

44. G. H. Hardy, 'What is geometry?', *Mathematical Gazette* 12 (1925), 309–16 = *Papers*, Vol. VII (1979), 519–26, and *Reader*, pp. 235–47. The 'curious imaginary curve' appeared in the *Mathematical Gazette* 4 (1907), 14 = *Papers*, Vol. VII (1979), 483.

45. G. H. Hardy, 'The case against the Mathematical Tripos', *Mathematical Gazette* 13 (1926), 61–71 = *Papers*, Vol. VII (1979), 527–37, or *Reader*, pp. 249–66.

46. This story appears in 'Obituary: George Pólya', *Bulletin of the London Mathematical Society* 19 (1987), 559–608, on pp. 561–2.

47. G. H. Hardy, 'Prolegomena to a chapter on inequalities', *Journal of the London Mathematical Society* 4 (1929), 61–78, and 5 (1930), 80 = *Papers*, Vol. II (1967), 471–89.

48. The colleague was Christopher Cox. (Courtesy of the Warden and Fellows of New College, Oxford; NCA Cox papers, PA/C1/6/4/268).

49. Hardy's obituaries of James Glaisher, David Hilbert, Edmund. Landau, and Gösta Mittag-Leffler, and some of his many book reviews, can be found in *Reader*, pp. 325–70.

50. Hardy's Cambridge Tracts in Mathematics and Mathematical Physics were: No. 2, *The Integration of Functions of a Single Variable* (1905); No. 12, *Orders of Infinity: The Infinitärcalcul of Paul Du Bois-Raymond* (1924); No. 18, *The General Theory of Dirichlet's Series* (with Marcel Riesz) (1915); and No. 38, *Fourier Series* (with W. W. Rogosinski) (1944).

51. G. H. Hardy, *A Course of Pure Mathematics*, Cambridge University Press (1908).

52. G. H. Hardy and E. M. Wright, *An Introduction to the Theory of Numbers*, Clarendon Press (1938); a revised and updated edition by R. Heath-Brown, J. Silverman, and A. Wiles was published in 2008.

53. See *Apology*, pp. 50–1.

54. G. H. Hardy, 'Mathematics', *Oxford Magazine* (5 June 1930), 819–21 = *Papers*, Vol. VII (1979), 607–9, or *Reader*, pp. 291–4.

55. See note 28.

56. Letter from G. H. Hardy to Oswald Veblen, 22 June 1931.

57. Hardy's manuscript on 'Round numbers' appears as the Frontispiece to *Papers*, Vol. VII.

58. *Apology*, p. 57; see also *Kanigel*, p. 369, or *Reader*, p. 81.

59. *Apology*, p. 148; see also *Kanigel*, p. 369.

60. Titchmarsh, *Notes*.

61. Letter from G. H. Hardy to Oswald Veblen, 24 August 1931. Louis J. Mordell was an American-born British number-theorist who studied in Cambridge and taught for many years in London and Manchester, before returning to Cambridge in 1945 as Hardy's successor as Sadleirian Professor.

62. From a Speech Day talk given by E. C. Titchmarsh at his former school, King Edward VII School, Sheffield, 1957.

63. See M. L. Cartwright, 'Edward Charles Titchmarsh', *Journal of the London Mathematical Society* 39/1 (1964), 544–65, on pp. 544–5 = *Biographical Memoirs of Fellows of the Royal Society* 10 (1964), 305–24, on pp. 305–6.

64. From Titchmarsh, *Notes*.

65. See Cartwright (note 63), on p. 546 (306).

66. From Titchmarsh, *Notes*.

67. E. C. Titchmarsh, *The Zeta-function of Riemann*, Cambridge Tracts in Mathematics and Mathematical Physics 26, Cambridge University Press (1930); *The Theory of the Riemann Zeta-function*, Clarendon Press (1951).

68. 'Prof. E. C. Titchmarsh. Mathematics at Oxford', Obituary in *The Times*, 19 January 1963, and the *New College Record*, 1962–63, on p. 2.

69. E. C. T. (Edward Thompson), 'E. C. Titchmarsh', *Oxford Magazine* (28 February 1963), 212–13.

70. As note 62.

71. Dr W. L. Ferrar, Principal of Hertford College, Oxford, in a letter to *The Times*, 23 January 1963 (p. 15), shortly after Titchmarsh's death.

72. E. C. Titchmarsh, *The Theory of Functions*, Clarendon Press (1932).

73. As note 71.

74. E. C. Titchmarsh, *Introduction to the Theory of Fourier Integrals*, Clarendon Press (1937); for the quotation see Cartwright (note 63), on p. 555 (315).

75. E. C. Titchmarsh, *Eigenfunction Expansions associated with Second-order Differential Equations*, Clarendon Press (Part I, 1946; Part II, 1958); the extract appears at the beginning of Part II, see Cartwright (note 62), on p. 555 (316).

76. E. C. Titchmarsh, *Mathematics for the General Reader*, Hutchinson's University Library (1948); Dover paperback, 2017.

77. E. C. Titchmarsh, 'The art of proof', *The Listener* (23 April 1959), 714–16.

78. Both tributes come from Titchmarsh, *Notes*.

79. See Edward Thompson (note 69).

CHAPTER 7

Further reading

An excellent biographical account of Michael Atiyah's life and works is
   Nigel Hitchin, 'Sir Michael Atiyah OM. 22 April 1929–11 January 2019', *Biographical Memoirs of Fellows of the Royal Society* 69 (September 2020), 9–35. [*Hitchin*]

There is also a chapter by Michael Atiyah on 'Some personal reminiscences' in *Oxford Figures*.

Several of the references listed below can be found in
   Michael Atiyah: Collected Works, 7 vols, Clarendon Press (1988–2014). [*Works*]

Notes and references

1. M. L. Cartwright, 'Edward Charles Titchmarsh', *Journal of the London Mathematical Society* 39/1 (1964), 544–65 = *Biographical Memoirs of Fellows of the Royal Society* 10 (1964), 305–24.

2. *Oxford Figures*, p. 328.

3. Speech given by Graeme Segal at a memorial event for Michael Atiyah.

4. Sir Michael Atiyah, OM, FRS, FRSE, 'Mathematics into the 20th century', *Bulletin of the London Mathematical Society* 34 (2002), 1–15, on p. 7 = *Works*, Vol 6, pp. 651–65, on p. 657.

5. Michael Atiyah, 'What is geometry?', The 1982 Presidential address of the Mathematical Association, *Mathematical Gazette* 66/437 (October 1982), 179–84, on pp. 183–4 = *Works*, Vol. 1, pp. 289–94, on pp. 293–4.

6. J. Meek, 'I'm a bit of a jack of all trades', *The Guardian*, Wednesday 21 April 2004.

7. Michael Francis Atiyah, 'John Arthur Todd, 23 August 1908–22 December 1994', *Biographical Memoirs of Fellows of the Royal Society* 42 (November 1996), 483–94, on p. 484, reprinted in the *Bulletin of the London Mathematical Society* 30 (1998), 305–16, on p. 306 = *Works*, Vol. 6, pp. 599–610, on p. 600.

8. Roberto Minio, 'An interview with Michael Atiyah', *Mathematical Intelligencer* 6/1 (1984), 9–19, on p. 10 = *Works*, Vol. 1, pp. 297–307, on p. 298.

9. Minio (note 8), on pp. 14–15 = *Works*, Vol. 1, on p. 302.

10. 'How Lily got to Girton and beyond' by Michael Atiyah, in the programme for 'Lily Atiyah, 8 Jan 1928–13 Mar 2018, A celebration of Lily's life and work', Playfair Library, Edinburgh, 20 July 2018.

11. See *Hitchin*, and the filmed interviews with Atiyah on *www.webofstories.com/play/michael. atiyah/7*.

12. 'World's great minds gather to celebrate Atiyah's birthday', *Edinburgh Herald*, 20 April 2009.

13. See note 3.

14. Personal communication from Jill Strang.

15. 'Speech on conferment of Antonio Feltrinelli Prize, 1984', pp. 183–8 on p. 188 = *Works*, Vol. 1, 311–16, on p. 316.

16. Michael Atiyah, 'A personal history', *Works*, Vol. 6, pp. 9–15, on p. 9.

17. Minio (note 8), on p. 17 = *Works*, Vol. 1, on p. 305.

18. See *Oxford Figures*, p. 327.

19. See *Oxford Figures*, p. 328.

20. See note 19.

21. From Michael Atiyah's commentary on 'General Papers 10–23' in *Works*, Vol. 1, p. 186.

22. Personal communication from Robin Wilson.

23. Minio (note 8), on p. 13 = *Works*, Vol. 1, on p. 301.

24. Atiyah (note 16), in *Works*, Vol. 6, on p. 14.

25. Michael Atiyah, 'On the work of Edward Witten', *Proceedings of the International Congress of Mathematicians, Kyoto, Japan* (1990), 31–5, on p. 31 = *Works*, Vol. 6, 209–13, on p. 209.

26. Edward Witten, 'Michael Atiyah and physics', in a memorial tribute, 'Memories of Sir Michael Atiyah' (ed. N. Hitchin), *Notices of the American Mathematical Society* 66/11 (December 2019), 1837–9, on p. 1839.

27. Feltrinelli prize (note 15), on p. 188 = *Works*, Vol. 1, on p. 316.

28. M. F. Atiyah, F.R.S., 'Bakerian lecture 1975: Global geometry', *Proceedings of the Royal Society (London), Series A*, 347 (1976), 291–9, on p. 292 = *Works*, Vol. 1, pp. 219–27, on p. 220.

29. Michael Atiyah, 'The unity of mathematics', Presidential address to the London Mathematical Society, 19 November 1976, *Bulletin of the London Mathematical Society* 10 (1978), 69–76, on p. 76 = *Works*, Vol. 1, pp. 279–86, on p. 286.

30. The extracts here are from Minio (note 8), on pp. 10, 12, 16, and 17 = *Works*, Vol. 1, on pp. 298, 300, 304, and 305.

31. Minio (note 8), on p. 10 = *Works* Vol. I, on p. 298.

32. Atiyah (Note 16), in *Works*, Vol. 6, on p. 10.

33. I. M. James (ed.), *History of Topology*, Elsevier Science, B.V., 1999. Also, Ioan James, *Remarkable Mathematicians: From Euler to von Neumann; Remarkable Physicists: From Galileo to Yukawa; Remarkable Biologists: From*

Ray to Hamilton; *Remarkable Engineers: From Riquet to Shannon*, Cambridge University Press (2002, 2004, 2009, and 2010).

34. Ioan James, *Driven to Innovate: A Century of Jewish Mathematicians and Physicists*, Peter Lang (2009); *Asperger's Syndrome and High Achievement: Some Very Remarkable People*, Jessica Kingsley Publishers (2005); *The Mind of the Mathematician* (with Michael Fitzgerald), Johns Hopkins University Press (2007).

35. Andrew Wiles, 'Modular elliptic curves and Fermat's Last Theorem' and Andrew Wiles, 'Modular elliptic curves and Fermat's Last Theorem' and Richard Taylor and Andrew Wiles, 'Ring-theoretic properties of certain Hecke algebras', *Annals of Mathematics* 141/3 (May 1995), 443–551 and 553–72.

36. Martin Raussen and Christian Skau, 'Interview with Michael Atiyah and Isadore Singer', *European Mathematical Society Newsletter* 53 (September 2004), 24–30, reproduced in *Notices of the American Mathematical Society* 52/2 (February 2005), 223–31.

CHAPTER 8

Notes and references

1. M. F. Atiyah, N. J. Hitchin, V. G. Drinfeld, and Yu. I. Manin, 'Construction of instantons', *Physics Letters A* 65/3 (1978), 185–7.

2. N. J. Hitchin, A. Karlhede, U. Lindström, and M. Roček, 'Hyperkähler metrics and supersymmetry', *Communications in Mathematical Physics* 108 (1987), 535–89.

NOTES ON CONTRIBUTORS

Philip Beeley is Research Fellow and Tutor in the Faculty of History and Fellow of Linacre College, Oxford. A former President of the British Society for the History of Mathematics, he is editor (with the late Christoph J. Scriba) of the multi-volume *Correspondence of John Wallis (1616–1703)*, published by OUP. His main research interests are the history of calculus, mathematical correspondence networks in early–modern Europe, the history of scientific institutions (especially the Royal Society and the Leopoldina), and mathematical reading practices in the 17th century. He is co-editor of *Reading Mathematics in Early Modern Europe* (2021) and *Beyond the Academy: The Practice of Mathematics 1600–1850* (OUP).

Allan Chapman teaches at the University of Oxford, where he is a member of the Faculty of History and of Wadham College, and is a former Visiting Professor in the History of Science at Gresham College. His interests lie mainly in the history of astronomy, with particular emphasis on the development of astronomical instruments and observatories. An enthusiastic popularizer of his subject, he has written biographies of many scientists, including Robert Hooke and Mary Somerville.

Keith Hannabuss graduated and completed a doctorate in mathematics at Oxford. A Fellow by Examination at Magdalen College, a Royal Society Visitor at the ETH in Zürich, and a Moore instructor at MIT, he returned to Balliol College as a Fellow and Tutor in Mathematics, teaching there for forty years until his retirement. His main research interest is in mathematical physics, and he is author of the text *Introduction to Quantum Theory* (OUP 1997), but he also has a longstanding interest in the history of mathematics, particularly of the 19th century.

Christopher Hollings is Departmental Lecturer in Mathematics and its History in the Oxford Mathematical Institute, and Clifford Norton Senior Research Fellow in the History of Mathematics at The Queen's College, Oxford. His research covers a range of aspects of the history of mathematics in the 19th and 20th centuries, including the development of abstract algebra, the International Congresses of Mathematicians, and the modern historiography of ancient Egyptian mathematics.

Frances Kirwan is the current Savilian Professor of Geometry at Oxford University. After studying mathematics at Clare College, Cambridge, she moved to Oxford for her DPhil degree, supervised by Michael Atiyah. She was a Junior Fellow at Harvard University from 1983 to 1985 before returning to Oxford as a Fellow by Examination at Magdalen College. She became a Tutorial Fellow in Mathematics at Balliol College with a Lecturership (and from 1994 a Readership) at the University from 1986 to 2017, before being appointed to the Savilian Chair. A Fellow of the Royal Society, she served on its Council from 2012 to 2015, was President of the London Mathematical Society from 2003 to 2005, chaired the UK Mathematics Trust from 2010 to 2016, and was awarded a DBE in 2014. She was awarded the Royal Society's Sylvester Medal in 2021.

Mark McCartney is Senior Lecturer in Mathematics at Ulster University and former President of the British Society for the History of Mathematics. His research in applied mathematics is centred around nonlinear dynamics, while his work in the history of mathematics and natural philosophy finds its focus in the 19th century. His most recent edited book (with Andrew Whitaker and Alastair Wood) is *George Gabriel Stokes: Life, Science and Faith* (OUP 2019).

Karen Hunger Parshall is Commonwealth Professor of History and Mathematics at the University of Virginia, USA, where she has served on the faculty since 1988. Her research has focused on the history of science and mathematics in America, the history of 19th- and 20th-century algebra, and the development of national mathematics research communities. A Guggenheim Fellow in 1996–97, an Inaugural Fellow of the American Mathematical Society in 2012, and winner of its 2018 Albert Leon Whiteman Prize, she was an invited speaker at the International Congress of Mathematicians in Zürich in 1994. Her latest book *The New Era in American Mathematics, 1920–1950* is published by Princeton University Press.

William Poole is Fellow and Tutor in English, Senior Tutor, and Fellow Librarian of New College, Oxford. He writes on intellectual, literary, and scientific history, and bibliography, and is co-editor of *The Library*. Recent books include *Milton and the Making of Paradise Lost* (2017), *John Fell's New Year Books* (2018), and *Thomas Hyde: Epistola de mensuris et ponderibus Serum seu Sinensium (1688)* (2021).

Benjamin Wardhaugh is a former Fellow of All Souls College, Oxford, and works on early modern mathematics. His interests include the mathematical music theories of the 17th century and the practices of mathematical reading and annotation. He is the author of several books about mathematics in the past, including (most recently) *The Book of Wonders*, a history of Euclid's *Elements* from antiquity to the present.

Robin Wilson is an Emeritus Professor of Pure Mathematics at the Open University, and of Geometry at Gresham College, London, and is a former Fellow of Keble College, Oxford University. A former President of the British Society for the History of Mathematics, he has written and edited around fifty books on mathematics and its history, including fifteen books for OUP. Involved with the popularization and communication of these subjects, he has received international awards for his 'outstanding expository writing' and for his outreach activities.

PICTURE SOURCES AND ACKNOWLEDGEMENTS

Many people and institutions have helped in the preparation of this book, and in a variety of ways. In particular, grateful thanks are due to the Weston Library (part of Oxford University's Bodleian Libraries), the British Society for the History of Mathematics, and New College, Oxford for hosting and funding the one-day meeting in November 2019 which led to the production of this volume.

The authors and editor are also grateful to the Bodleian Libraries for granting permission to reproduce manuscript images and copyright material and also to the following Oxford University institutions for support, help, and advice: the Mathematical Institute, the Oxford University Museum, the History of Science Museum, and Balliol College, New College, Pembroke College, and The Queen's College. We also thank St John's College, Cambridge and Trinity College, Cambridge, the Universities of Aberdeen and Warwick, the London Mathematical Society, the Mathematische Forschungsinstitut Oberwolfach, and the Abel Prize and Shaw Prize organizations.

We also wish to thank the following individuals: Jennifer Andrews, June Barrow-Green, Béla Bollobás, Ken Choi, Howard Colquhoun, Andrew Coulson, Adam Crothers, Danielle Czerkaszyn, Linsey Darby, Sue Dopson, Ellen Embleton, Howard Emmens, Alain Enthoven, Mordechai Feingold, Nina Fowler, Daniel Fried, Bethany Hamblen, Nigel Hitchin, Amanda Ingram, Ioan James, Leonard Klosinski, Jeannie Lee, Petra Lein, Dyrol Lumbard, Roger Mallion, Sally Mullins, Susan Oakes, Michael Riordan, Heather Riser, Håkon Sandbakken, Fenny Smith, Jonathan Smith, Jim Stimpert, Richard Taylor, Jennifer Thorp, and Miles Young, and at Oxford University Press, Charlie Bath, Dan Taber, Katherine Ward, Sivanesan Ashok Kumar and Integra Software services.

Every effort has been made to trace all copyright holders, but if any have been inadvertently overlooked, the publishers will be pleased to make the necessary arrangements at the earliest opportunity.

FRONTISPIECE

0.0   Sir Henry Savile: Wikimedia Commons.

CHAPTER 1: SIR HENRY SAVILE AND THE EARLY PROFESSORS

1.1   Thomas Allen, etching by J. Bretherton, in R. T. Gunther, *Early Science in Oxford*, Vol. 11, Clarendon Press (1937), 286.

         Erasmus Williams: photograph by John Fauvel.

1.2   Savile's lectures on new astronomy: courtesy of the Bodleian Library, Savile, MS 31.

1.4   'Cista mathematica', in R. T. Gunther, *Early Science in Oxford*, Vol. 1, Clarendon Press (1923), opposite p. 122.

         Model with Platonic solids, in R. T. Gunther, *Early Science in Oxford*, Vol. 1, Clarendon Press (1923), opposite p. 122; the model is housed in the History of Science Museum, Oxford.

1.5   Sir Henry Savile, *Lectures on Euclid* (Oxford, 1621), Bodleian Library Bod 1393 d.23: Wikimedia Commons.

1.6   John Napier, *Mirifici Logarithmorum Canonis Descriptio* (Edinburgh, 1614).

1.7   Henry Briggs, *Logarithmorum Chilias Prima* (1617).

         Henry Briggs, *Arithmetica Logarithmica* (London, 1624).

1.8   Henry Briggs map: Wikimedia Commons.

1.10   John Bainbridge, in R. Poole, *Catalogue of Portraits in the Possession of the University, Colleges, City, and County of Oxford*, Vol. 2 (Oxford, 1912), opposite p. 44.

1.11   Bainbridge's manuscript of Ulugh Beg: courtesy of the Bodleian Library, MS Savile 46.

         John Bainbridge, *Canicularia* (Oxford, 1648): Wikimedia Commons.

1.13   Turner's *Encyclopaedia seu Orbis Literarum*, Oxford University Statuta Selecta (1638).

1.14   John Greaves, in R. T. Gunther, *Early Science in Oxford*, Vol. 11, Clarendon Press (1937), 48.

         Diagram from *Pyramidographia* (London, 1646): Wikimedia Commons.

CHAPTER 2: JOHN WALLIS

2.0   Portrait of John Wallis: Wikimedia Commons.

2.1   1649 pamphlet: *A Serious and Faithfull Representation of the Judgements of the Ministers of the Gospell Within the Province of London, Jan. 18, 1649.*

2.2   Collection of letters: Bodleian MS e Musaeo 203.

2.3   Portrait of William Oughtred by Wenceslas Hollar: Wikimedia Commons.

         William Oughtred, *Clavis Mathematicae*, 3rd edn (Oxford, 1652).

2.4   John Wallis, *De Sectionibus Conicis Tractatus* (Oxford, 1655), 4–5.

2.5   John Wallis, *Arithmetica Infinitorum* (Oxford, 1656), title page and Proposition 121.

2.6   John Wallis, *Arithmetica Infinitorum*, table and formula from Propositions 189 and 191.

2.7   John Wallis, *In Elementa Euclidis*, Bodleian Library, MS Don. d. 45.

2.8   Wallis's inaugural lecture, in *Opera Mathematica* (Oxford, 1693–99).

         John Wallis, *De Cycloide et Corporibus Inde Genitis* (Oxford, 1659).

2.9   John Wallis, *Elenchus Geometriae Hobbianae* (Oxford, 1655).

         Thomas Hobbes's *Six Lessons to the Professors of the Mathematiques* (London, 1656).

2.10   Wallis's manuscript on a theorem of Fermat, Bodleian Library, MS Don. C.49, f. 14r–15v.

2.11   John Wallis's *Adversus Marci Meibomii* (Oxford, 1657).

2.12   John Wallis's *A Discourse of Gravity and Gravitation* (London, 1674).

         John Wallis, 'A Summary Account . . . of the General Laws of Motion', *Philosophical Transactions* 43 (11 January 1668/9), 864–6.

John Wallis, *Mechanica* (London, 1670–71).

2.13 John Wallis, *A Treatise of Algebra, both Historical and Practical* (London, 1685), title page and p. 266.

2.14 John Wallis, *Opera Mathematica* (Oxford, 1693–99).

2.15 John Wallis, 'Reasons Shewing the Consistency of the Place of Custos Archivorum with that of a Savilian Professor' (Oxford, 1658).

2.16 John Wallis, *Theological Discourses* (London, 1692).

2.17 Letter from John Wallis to Thomas Smith, 29 January 1696/7, Bodleian Library, MS Smith 31, p. 38.

CHAPTER 3: A CENTURY OF ASTRONOMERS: FROM HALLEY TO RIGAUD

3.0 Edmond Halley's *Apollonii Pergaei Conicorum Libri Octo* (Oxford, 1710), frontispiece.

3.1 Portrait of 'young Halley' by Thomas Murray: Wikimedia Commons.

3.2 Halley's appearance in The Queen's College Entrance Book: by kind permission of the Provost, Fellows and Scholars of The Queen's College, Oxford.

3.3 Halley's 'Robur Carolinum', from J. Hevelius's *Uranigraphia*: Wikimedia Commons.

3.4 Halley's chart of magnetic variation: Wikimedia Commons.

3.5 Path of the total solar eclipse of 1715: Wikimedia Commons.

3.6 E. Halley, 'Astronomiae cometicae synopsis', *Philosophical Transactions* 24 (1704–5), 1882.

3.7 E. Halley, *Apollonii de Sectione Rationis* (Oxford, 1706), title page.
    E. Halley, *Apolonii Pergaei Conicorum Libro Octo* (Oxford, 1710), title page.

3.8 E. Halley, *Menelai Sphaericorum Libri III* (Oxford, 1758), Prop. XI.: Bodleian Library, Rigaud e.70 (4).

3.9 Diagrams from Halley's work: *Philosophical Transactions* 16 (1686–92), after 402.

3.10 House in New College Lane, in R. T. Gunther, *Early Science in Oxford*, Vol. 2, Clarendon Press (1923), 83.

3.11 Mezzotint of Edmond Halley, by J. Faber after T. Murray: Wikimedia Commons.

3.12 Portrait of Nathaniel Bliss: Wikimedia Commons.
    Drawing of Bliss, in R. T. Gunther, *Early Science in Oxford*, Vol. 11, Clarendon Press (1937), 276, and Wikimedia Commons.

3.13 Bliss advertisements: courtesy of the History of Science Museum, Oxford, and Robin Wilson's article, 'Ignorance of Bliss', *The Pembrokian*, No. 34 (2010), 15–17, by kind permission of the Master, Fellows, and Scholars of Pembroke College, Oxford.

3.14 Joseph Betts's engraving of the solar eclipse, courtesy of the History of Science Museum, Oxford.

3.15 J. Smith, *Observations on the Use and Abuse of the Cheltenham Waters* (Cheltenham, 1796), title page.

3.16 Memorial to David Gregory, photograph by John Fauvel.

3.17 Gregory's inaugural lecture: courtesy of the University of Aberdeen, MS 2206.8.

3.18 J. T. Desaguliers, *Course of Experimental Philosophy* (London, 1763), image 528.

3.19 Portrait of James Bradley: Wikimedia Commons.

3.20 Course of lectures: courtesy of the History of Science Museum, Oxford.

3.21   A. Robertson, *Elements of Conic Sections* (Oxford, 1818), title page.

3.22   Stephen Peter Rigaud, *On the Arenarius of Archimedes*, Oxford (1837), title page.
        Silhouette of Rigaud, in R. Poole, *Catalogue of Portraits in the Possession of the University, Colleges, City, and County of Oxford*, Vol. 2 (Oxford, 1912), opposite p. 44.

3.23   Meridian circle. in A. D. Thackeray, *The Radcliffe Observatory 1772–1972*, The Radcliffe Trust (Oxford, 1972), Plate 14.

CHAPTER 4: BADEN POWELL AND HENRY SMITH

4.0   Henry Smith, carte de visite: courtesy of Vanda Morton.

4.1   Tables supplied by Keith Hannabuss.

4.2   Baden Powell: Wikimedia Commons, and R. T. Gunther, *Early Science in Oxford*, Vol. XI, Clarendon Press (1937), opposite p. 120.

4.3   Baden Powell, *The Present State and Future Prospects of Mathematical and Physical Studies at the University of Oxford*, Oxford (1832), title page.
        Baden Powell, *History of Natural Philosophy from the Earliest Period to the Present Time*, Longman (1837), title page.

4.4   William Donkin: Wikimedia Commons.
        Bartholomew Price: by kind permission of the Master, Fellows and Scholars of Pembroke College, Oxford, catalogue number PMB/M/3/2/5/14.

4.5   Baden Powell Royal Institution lecture, *Illustrated London News* (17 May 1851).

4.6   British Association meeting in 1847, *Illustrated London News* (3 July 1847).

4.7   George Gabriel Stokes: Wikimedia Commons.

4.8   University Museum under construction: courtesy of the Oxford University Museum of Natural History.
        Proposed design: *The Builder* (7 July 1855).
        Museum interior, *Illustrated London News* (6 October 1860).
        University Museum in 1860, *Oxford University Almanac* (1860).

4.9   Augustus De Morgan and George Boole: Wikimedia Commons.

4.10   Henry John Stephen Smith: courtesy of the London Mathematical Society.
        Caricature of Henry Smith: courtesy of the Bodleian Library, GA. Oxon. 414 4.

4.11   Maskelyne photographs of Baden Powell and Henry Smith: courtesy of Vanda Morton.

4.12   H. J. Stephen Smith M.A., 'Report on the Theory of Numbers, Part I', *Report of the British Association for 1859*, p. 228 = *Collected Mathematical Papers of Henry Smith*, Vol. 1 (1894), 38.

4.13   William Spottiswoode: Wikimedia Commons.
        James Glaisher: Wikimedia Commons, or courtesy of the London Mathematical Society.

4.14   University Museum's Keeper's House and bust of Henry Smith: courtesy of the Oxford University Museum of Natural History.

4.15   Charles Hermite, *Oeuvres de Charles Hermite*, Vol. II (1908), frontispiece.
        Felix Klein and Ferdinand Lindemann: Wikimedia Commons.

4.16   Smith's construction with $n = 3$, image supplied by Keith Hannabuss.

4.17   X-ray structure, and NMR simulation based on Smith construction: images supplied by Howard Colquhoun.

4.18   *The Collected Mathematical Papers of Henry John Stephen Smith M.A., F.R.S.*, Vol. 1, Oxford
        University Press (1894), frontispiece.
        Hermann Minkowski: Wikimedia Commons.

CHAPTER 5: JAMES JOSEPH SYLVESTER

5.0   Sylvester engraving by G. J. Stodart, from a photograph by Messrs. I. Stillard & Co.: Wikimedia
        Commons.
5.1   Sylvester portrait by George Patton: courtesy of Alain Enthoven.
5.2   'The Student': Special Collections, University of Virginia Library.
5.3   Arthur Cayley: Wikimedia Commons.
5.4   Mathematical Seminary: Ferdinand Hamburger Archives of The Johns Hopkins University.
5.5   Letter from Sylvester to Cayley: by permission of the Master and fellows of St John's College,
        Cambridge.
5.6   Sylvester's sonnet: by permission of the Warden and Scholars of New College Oxford.
5.7   Sylvester's paper on reciprocants: *American Journal of Mathematics* 9(1) (October 1886), 1.
5.8   Brill's model of surface: courtesy of the Mathematical Institute, University of Oxford.
        Sylvester announcement: *Oxford University Gazette*, 15 October 1886.
5.9   Edwin Bailey Elliott: courtesy of the London Mathematical Society.
5.10   Oxford Mathematical Society: courtesy of the Oxford Mathematical Institute.
5.11   Sylvester portrait by Alfred E. Emslie: Wikimedia Commons.
5.12   William Esson: courtesy of the London Mathematical Society.

CHAPTER 6: G. H. HARDY AND E.C. TITCHMARSH

6.0   G. H. Hardy at New College: courtesy of the Master and Fellows of Trinity College, Cambridge.
6.1   Bertrand Russell: Wikimedia Commons.
        Srinivasa Ramanujan: Wikimedia Commons.
        J. E. Littlewood: The George Pólya Collection and Leonard Klosiniski.
6.2   Inaugural lecture: *G. H. Hardy, Some Famous Problems of the Theory of Numbers and in particular
        Waring's Problem*, Clarendon Press (1920).
6.3   Engraving of New College: by permission of the Warden and Scholars of New College Oxford,
        NCA 5597.
6.4   Augustus Love: courtesy of the London Mathematical Society.
        A. L. Dixon, Source unknown.
6.5   G. H. Hardy and Mary Cartwright: courtesy of the family of Professor Robert A. Rankin.
6.6   Hardy and Littlewood: courtesy of the Master and Fellows of Trinity College, Cambridge, and
        Béla Bollobás.
6.7   Lecture list. *Oxford University Gazette*, Michaelmas Term, 1920.
6.8   Hardy's announcement of lectures, *Oxford University Gazette*, Michaelmas Term, 1920.
6.9   Lecture notes by E. H. Linfoot: courtesy of Béla Bollobás.
6.10   Hardy and Pólya in Oxford: The George Pólya Collection and Leonard Klosiniski.
        Hardy, Littlewood, Pólya, *Inequalities*, Cambridge University Press (1934).

6.11 Hardy's cricket team. *Daily Mirror* (11 August 1926).

6.12 'A curious imaginary curve'. *Mathematical Gazette* 4 (1907), 14.

6.13 G. H. Hardy, *A Mathematician's Apology*, 2nd edn, Cambridge University Press (1967).

6.14 *Quarterly Journal of Mathematics* 1 (1930), contents page.

6.15 Hardy at New College in 1930, in *The Mathematical Intelligencer* 15/3 (1993), 4.

6.16 Einstein's visit to Oxford, *Illustrated London News* (16 May 1931), 832.
Blackboard from Einstein lecture: courtesy of the History of Science Museum, Oxford.

6.17 Notes on 'Round Numbers': *Collected Papers of G. H. Hardy*, Vol. 7, Clarendon Press (1979), frontispiece.

6.18 E. C. Titchmarsh: courtesy of the Titchmarsh family.

6.19 Titchmarsh's letter of appointment: by permission of the Warden and Scholars of New College Oxford.

6.20 Letter from Hardy to Titchmarsh: courtesy of Béla Bollobás.

6.21 Titchmarsh, Coulson, and Temple: courtesy of the Coulson family.

6.22 E. C. Titchmarsh, *The Theory of Functions*, Clarendon Press (2nd edn, 1939) and *Introduction to the Theory of Fourier Integrals*, Clarendon Press (2nd edn, 1948).

CHAPTER 7: FROM MICHAEL ATIYAH TO THE 21ST CENTURY

7.0 Michael Atiyah in the Mathematical Institute: courtesy of the London Mathematical Society.

7.1 John A. Todd: courtesy of the Godfrey Argent Studio.
William Hodge: courtesy of the London Mathematical Society.

7.2 Lily Atiyah: source unknown.
Mary Cartwright: Wikimedia Commons.

7.3 Michael Atiyah and Fritz Hirzebruch: Wikimedia Commons.

7.4 Isadore Singer: Wikimedia Commons.
Raoul Bott: Wikimedia Commons.

7.5 Henry Whitehead: source unknown.

7.6 Michael Atiyah lecturing: Wikimedia Commons.
Michael Atiyah lecturing, from *The Illustrated History of Oxford University* (ed. John Prest), Oxford University Press (1993), 324.

7.7 Graeme Segal (left) and Ed Witten: Wikimedia Commons.

7.8 Three Fields Medallists, courtesy of the Mathematical Institute, University of Oxford.
Abel Prize ceremony: photograph by Knut Falch/Scanpix.

7.9 Ioan James: courtesy of Ioan James and the Mathematical Institute, University of Oxford.

7.10 I. M. James (ed.), *Handbook of Algebraic Topology*, North-Holland (1995), and *Topological and Uniform Spaces*, Springer-Verlag (1987).

7.11 Richard Taylor: Wikimedia Commons.

7.12 Nigel Hitchin: Wikimedia Commons.
Frances Kirwan: courtesy of Frances Kirwan.

7.13 *Frances Kirwan*, Oil on canvas, 82 × 98 cm, Nina Mae Fowler, 2019, Collection of Balliol College, Oxford.

CHAPTER 8: INTERVIEW WITH NIGEL HITCHIN

8.0   Nigel Hitchin conferences: courtesy of the Mathematical Institute, University of Oxford.

8.1   Nigel Hitchin as a student: courtesy of Nigel Hitchin.

8.2   Hitchin, Atiyah, and Yau: Author: Dirk Ferus. Source Archives of the Mathematische Forschungsinstitut Oberwolfach.

8.3   Honorary degree: courtesy of the University of Warwick.

8.4   Shaw Prize for Mathematical Sciences: courtesy of The Shaw Prize Foundation and The Hong Kong University of Science and Technology.

8.5   Nigel Hitchin at work: courtesy of Nigel Hitchin.

INDEX OF NAMES

Q

Quillen, Daniel 146, 198

R

Ramanujan, Srinivasa 147–8, 150, 160–1, 172
Rawnsley, John 215
Riesz, Marcel 166
Rigaud, Stephen 55, 84, 87, 90–1, 93
Robbins, Lionel 152
Robertson, Abraham 55, 74, 84–90, 93
Roberval, Gilles Personne de 30, 41
Rogers, Leonard 137
Rogosinski, W. W. 166
Russell, Bertrand 147–8
Russell, John Wellesley 108, 152, 158, 173–4

S

Saccheri, Girolamo 40
Salmon, George 132–3
Sampson, C. H. 149, 152, 158
Savile, Sir Henry 2ff, 19–20, 22, 27, 37, 40, 42, 54, 61, 65, 69
Scaliger, J. J. 23
Schoute, Pieter 134
Scot, Thomas 49
Segal, Graeme 184, 190, 195, 214, 217
Sewell, James 130
Singer, Isadore 190–1, 194–5, 198, 213, 219
Smith, Eleanor 110, 118
Smith, Henry J. S. 92–4, 99–100, 105ff, 121–2, 128, 145, 150
Smith, John 34

Smith, John 55, 74–6, 84, 93
Smith, Thomas 52–3
Snow, C. P. 146, 167
Spooner, William 149, 151
Spottiswoode, William 99, 112–13, 129
Stanley, Gertrude 155
Steer, Brian 203, 211–2
Stokes, George Gabriel 101–2, 108
Swift, Jonathan 63, 79
Sylvester, James Joseph 94, 109, 120ff, 145–6, 150, 154, 164, 204

T

Tait, Campbell 103
Taylor, Richard 183, 201–3, 217
Temple, George 177
Thomas, Lyn 210
Thompson, C. H. 152, 158
Thompson, Edward 178, 211
Thompson, John 216
Titchmarsh, Edward C. 145, 155–6, 163, 168–9, 172ff, 183–4, 188–9
Titchmarsh, Kathleen 175
Todd, John A. 186–7
Tooke, Mary 60
Torelli, Giuseppe 85, 90
Torricelli, Evengelista 35–6
Turner, Herbert H. 149, 152, 154, 157–8, 164
Turner, Peter 23–7

U

Ulugh Beg 22–3
Ussher, James 14–15, 20, 22, 26–7

V

Veblen, Oswald 165, 168–70, 173
Vere, Lady Mary 32
Viazovska, Maryna 115
Vlacq, Adriaen 16

W

Waller, Richard 68
Wallis, John 23, 28ff, 55, 57, 61, 63–5, 79
Ward, Richard 194
Ward, Seth 7, 41–2
Waring, Edward 150
Warner, Walter 41
Weyl, Hermann 187
White, Thomas 8
Whitehead, Henry 146, 178, 192, 199–200
Wilberforce, Samuel 103
Wiles, Andrew 201–2
Wilkins, John 32, 39
Williams, Erasmus 5–6
Williamson, Sir Joseph 58
Witten, Edward 194–5
Wood, Anthony 25, 56
Wren, Christopher 46, 56
Wright, Edward 16
Wright, Sir Edward M. 155, 167

Y

Yang, C. N. 215
Yau, Shing-Tung 214
Young, William Henry 149

Z

Zagier, Don 211